Applications of Walsh and Related Functions

With an Introduction to Sequency Theory

1984

MICROELECTRONICS AND SIGNAL PROCESSING

Series editors: **P. G. Farrell,** University of Manchester, U.K.
J. R. Forrest, University College London, U.K.

About this series:
The topic of microelectronics can no longer be treated in isolation from its prime application in the processing of all types of information-bearing signals. The relative importance of various processing functions will determine the future course of developments in microelectronics. Many signal processing concepts, from data manipulation to more mathematical operations such as correlation, convolution and Fourier transformation, are now readily realizable in microelectronic form. This new series aims to satisfy a demand for comprehensive and immediately useful volumes linking the microelectronic technology and its applications.

Key features of the series are:
• Coverage ranging from the basic semiconductor processing of microelectronic circuits to developments in microprocessor systems or VSLI architecture and the newest techniques in image and optical signal processing.
• Emphasis on technology, with a blend of theory and practice intended for a wide readership.
• Exposition of the fundamental theme of signal processing; namely, any aspect of what happens to an electronic (or acoustic or optical) signal between the basic sensor which gathers it and the final output interface to the user.

1. *Microprocessor Systems and Their Application to Signal Processing:* C. K. YUEN, K. G. BEAUCHAMP, and G. P. S. ROBINSON
2. *Applications of Walsh and Related Functions:* K. G. BEAUCHAMP

In preparation

3. *Secure Speech Communications:* H. J. BEKER and F. C. PIPER

WM Sit tus eve
to guess

Tables   1:1   6
1:1 26   1:2   8
1:2 28   1:3   9
1:3   1:4   20
1:4 32   1:5   12
           1:6   13
           1:7   16
           1:8   19
           1:9   22
           1:10 23
           1:11 24 26/28
           1:12/1:12 29
           1:13   32

t 1.5
t 1.6 / 1.7
t 1.8
1-15
Refs.

2y
35
3y

Applications of Walsh and Related Functions

With an Introduction to Sequency Theory

1984

K. G. BEAUCHAMP

University of Lancaster
Lancaster, England

1984

ACADEMIC PRESS

(Harcourt Brace Jovanovich, Publishers)
London   Orlando   San Diego   New York
Toronto   Montreal   Sydney   Tokyo

COPYRIGHT © 1984, BY ACADEMIC PRESS INC. (LONDON) LTD.
ALL RIGHTS RESERVED.
NO PART OF THIS PUBLICATION MAY BE REPRODUCED OR
TRANSMITTED IN ANY FORM OR BY ANY MEANS, ELECTRONIC
OR MECHANICAL, INCLUDING PHOTOCOPY, RECORDING, OR ANY
INFORMATION STORAGE AND RETRIEVAL SYSTEM, WITHOUT
PERMISSION IN WRITING FROM THE PUBLISHER.

ACADEMIC PRESS INC. (LONDON) LTD.
24-28 Oval Road,
London NW1 7DX

United States Edition published by
ACADEMIC PRESS, INC.
Orlando, Florida 32887

British Library Cataloguing in Publication Data

Beauchamp, K. G.
  Applications of Walsh and related functions.
  –(Microelectronics and signal processing)
  1. Walsh functions
  I. Title   II. Series
  515'.55   QA404.5

Library of Congress Cataloging in Publication Data

Beauchamp, K. G.
  Applications of Walsh and related functions, with an
introduction to sequency theory.

  (Microelectronics and signal processing)
  Bibliography: p.
  Includes index.
    1. Signal processing.   2. Sequency theory.   3. Walsh
function.   I. Title.   II. Series.
TK5102.5.B346   1984   621.3'01'51555   84-6389
ISBN 0-12-084180-0 (alk. paper)

PRINTED IN THE UNITED STATES OF AMERICA

84 85 86 87   9 8 7 6 5 4 3 2 1

Contents

Preface ix
Abbreviations and Symbols xiii

PART 1 THEORY AND PRACTICE

Chapter 1 The Sequency Functions

1.1	Introduction	3
1.2	Orthogonality	4
1.3	The Walsh function series	15
1.4	The Haar function series	38
1.5	Mixed function series	40
1.6	Discrete sampled functions	44
	References	45

Chapter 2 Transformation

2.1	Introduction	48
2.2	The discrete Walsh transform	49
2.3	Fast Walsh transform algorithms	58
2.4	The discrete Haar transform	75
2.5	The discrete slant transform	82
2.6	Shift-invariant transformation	86
2.7	Transform conversion	87
2.8	Two-dimensional transformation	93
	References	95

Chapter 3 Analysis and Processing

3.1	Introduction	98
3.2	Correlation and convolution	99
3.3	Spectral analysis	103
3.4	Digital filtering	118
3.5	Waveform synthesis	129
	References	133

Chapter 4 Hardware Techniques

4.1	Introduction	137
4.2	Walsh function generators	138
4.3	Transformation	145
4.4	LSI application	158
	References	169

PART 2 APPLICATIONS

Chapter 5 Signal Processing

5.1	Introduction	175
5.2	Spectroscopy	176
5.3	Speech processing	177
5.4	Medical applications	179
5.5	Seismology	188
5.6	Non-linear applications	194
	References	197

Chapter 6 Image Processing

6.1	Introduction	201
6.2	Image compression	204
6.3	Image enhancement and restoration	214
6.4	Pattern recognition	220
	References	227

Chapter 7 Communications

7.1	Introduction	231
7.2	Multiplexing	232
7.3	Coding	240
7.4	Non-sinusoidal electromagnetic radiation	246
	References	260

Chapter 8 Logical Design and Analysis

8.1	Introduction	264
8.2	Rademacher–Walsh ordering	265
8.3	Synthesis of digital networks	268
8.4	Minimisation of logic functions	271
8.5	Fault diagnosis	281
	References	291

Selected List of Additional References   295
Index   301

Preface

Some years ago the writer had occasion to survey the range and feasibility for computer applications of Walsh functions. This later formed the material of a book published in 1975 in Academic Press's Techniques of Physics series under the title "Walsh Functions and Their Applications". Now, almost a decade later, it is pertinent to consider the subject area again and to record the progress and development that have taken place in the intervening period.

A number of these developments have been quite extensive, and in two areas, Boolean logic analysis and non-sinusoidal communication, substantial research fields have opened up which enable new techniques to be applied to the solution of problems previously inviolate to other methods of attack.

An overall impression of the work of the preceding decade is that of a broadening in the range of applicability together with a wider availability of different orthogonal transformations relevant to signal processing and communications. Not only are the several alternative orderings of the Walsh function seen to provide their own particular solution to a varied range of problems, but the specific features of Haar, slant, hybrid, cosine, sine and block series are being recognised as part of a set of processing tools now available to match against the characteristics of the problem. The availability of fast transformation algorithms for all of these functions and the use of the microprocessor as a system-processing component are two of the reasons for these developments.

It is not possible in a book of modest size to consider in any depth all the very many applications of Walsh and related functions that have been demonstrated throughout the physical sciences in recent years. Some quite important applications are described at a fairly elementary level simply because to do otherwise would result in the exclusion of several other equally important subjects. Others are omitted, not because of their lack of relevance but because their description would entail a mathematical treatment too extensive to be pursued in a book concerned with an overall view. One example lies in the use of sequency functions for systems and control, and another in the valuable contributions that have been made to dyadic theory and logic analysis in various countries during the past decade.

Instead a broad treatment of the main lines of development has been attempted, with the emphasis on understanding the principles involved. It is hoped that sufficient detail is included together with comparison of alternative methods so that the reader can assess the relevance to his own problems. To this end a considerable list of references is given at the end of each chapter and a list of further references and bibliographies, arranged by subject matter, is included at the end of this book.

The book consists of two parts. The first takes the form of a tutorial in sequency theory (Chapters 1–3) and gives the background essential for understanding the applications part which follows. Chapter 4 forms a bridging 'hardware' chapter between the earlier theoretical chapters and the application chapters (5–8).

Chapters 5 and 6 are concerned with signal processing in one and two dimensions. In this latter area the pace of development has increased in recent years, with sequency methods playing a significant role. It is likely that in the newer areas of robotic vision and satellite surveillance significant further progress will be made.

Chapter 7 describes applications in communications, and it is interesting to note the considerable progress that has been made recently in non-sinusoidal communications and radar, which has led to several commercial devices now becoming available.

Finally, Chapter 8 attempts to summarise another quite new field for analysis which is of considerable importance in the design and testing of integrated logic systems and is already producing significant practical results.

Selection and assembly of material for this book have relied considerably on the help given by very many people and organisations.

The writer would like to express his particular thanks for the assistance given by Professor H. Harmuth of the Catholic University of America and by Dr. S. Hurst of the University of Bath. Appreciation is also expressed to the following who have contributed in various ways to this book: Professor P. Besslich of Bremen University, Dr. B. Durgen of the University of Vermont,

Dr. W. Chen of the U.S. Army Topographical Laboratory, Dr. M. Hussain of Kuwait, Professor M. Karpovsky of Boston University, Professor R. Kitai of McMaster University, Mr. C. Nicol of British Telecomm, Professor R. Redinbo of Rensselaer Polytechnic Institute and Dr. C. Yuen of the University of Hong Kong. Finally, acknowledgement is extended to the Royal Society for travel support and to the Institution of Electrical Engineers for library assistance.

Lancaster   K. G. BEAUCHAMP

Abbreviations and Symbols

A–D	Analog-to-digital
CAD	Computer-aided design
CAL	Directly symmetrical Walsh function
CCD	Charge-coupled device
CMOS	Complementary metal–oxide–silicon
cos	Cosine
CT	Cosine transform
C–T	Cooley–Tukey
D–A	Digital-to-analog
DCT	Discrete cosine transform
DFT	Discrete Fourier transform
DHT	Discrete Haar transform
DMA	Direct memory access
DPCM	Differential pulse-coded modulation
DSM	Digital sequency multiplex
DST	Discrete sine transform
DWT	Discrete Walsh transform
ECD	Electrocardiograph
ECL	Emitter-coupled logic
EEG	Electroencephalograph
EPROM	Erasable programmable read-only memory
exp	Exponential
FCT	Fast cosine transform
FDM	Frequency division multiplex

FFT	Fast Fourier transform
FHT	Fast Haar transform
FST	Fast slant transform
FT	Fourier transform
FWT	Fast Walsh transform
HAD	Hadamard-ordered Walsh function
HAR	Haar function
HAW	Hadamard–Walsh function
HT	Haar transform
Hz	Hertz (cycles per second)
KLT	Karhunen–Loève transform
LSI	Large-scale integration
MSE	Mean-square error
MUX	Multiplexer
PAL	Paley-ordered Walsh function
PCM	Pulse-coded modulation
pel	Picture element
PLA	Programmable logic array
PROM	Programmable read-only memory
PSF	Point-spread function
RAD	Rademacher function
RAM	Random access memory
R–M	Reed–Muller
ROM	Read-only memory
SAL	Inversely symmetrical Walsh function
SAW	Surface acoustic wave
SDM	Sequency division multiplex
sin	Sine
SLA	Slant function
SLT	Slant transform
ST	Sine transform
TDM	Time division multiplex
VLSI	Very large-scale integration
WAL	Sequency-ordered Walsh function
WHT	Walsh–Hadamard transform
WT	Walsh transform
Zps	Sequency (zero crossings per second)
a_n	Fourier spectral coefficient
A	Unitary matrix
b	Binary digit
b_n	Fourier spectral coefficient
B	Bandwidth
BW	Besslich Rademacher–Walsh matrix

Abbreviations and Symbols

c	Phase velocity, velocity of light (3×10^8 m/s)
C	Channel capacity (bits per second)
C	Covariance matrix
$C(i, t)$	A set of orthogonal signal carriers
$CT(f)$	Discrete cosine transform
D	Diagonal matrix
e	2.71828
f	Frequency
f_c	Clock frequency
$f(t)$	Function of t
$f(x)$	Boolean logic function
F	As $F(x)$ with logic values 0 and 1 replaced by $+1$ and -1
$\overline{F}$	As $F(x)$ with logic values 0 and 1 replaced by -1 and $+1$
$F(\omega)$	Filter response
$F(x)$	Binary vertices of a truth table
F	Fourier transform matrix
G	Filter weights matrix
h_k	Weighting coefficient
H	Hadamard matrix
Ha	Haar matrix
i	Current; series coefficient
$i(x, y)$	Input image spatial domain
$i(\omega)$	Input image frequency domain
$\mathbf{I}_m$	Identity matrix
$\text{Im}(k)$	Imaginary value of k
j	$\sqrt{-1}$; series coefficient
k	Channel, constant
K	Kernel, filter weight, constant
m	Minterms of a logic function
n	Ordering number
N	Number of terms equalling 2^p
$o(x, y)$	Output image in spatial domain
$o(\omega)$	Output image in frequency domain
p	$\text{Log}_2 N$
$P_{aw}(k)$	Averaged Walsh power spectral coefficient
$P_F(k)$	Fourier power spectral coefficient
$P_H(k)$	Haar power spectral coefficient
P_N	Noise power
P_S	Signal power
$P_W(k)$	Walsh power spectral coefficient
P	Permutation matrix
r	Rademacher–Walsh series
R_i	Rademacher–Walsh spectral coefficients

$\text{Re}(k)$	Real value of k
$R_F(\tau)$	Correlation coefficient in real time
$R_W(\tau)$	Correlation coefficient in dyadic time
RW	Rademacher–Walsh matrix
s_i	Discrete signal
$\hat{s}_i$	Estimated discrete signal
s_{ij}	Two-dimensional discrete signal
$S(t)$	An orthogonal series
$ST(f)$	Discrete sine transform
t, T	Time
T	Threshold
v	Velocity (meters per second)
W	Walsh matrix
x_i	Sampled function of time; Boolean logic value
$\overline{x}_i$	Complemented Boolean logic value
$x_i(b)$	Binary function of time
$\mathbf{x}_{ij}$	Two-dimensional image matrix
$x(t)$	Continuous function of time
$X_c(k)$	CAL transform coefficient for x_i
$X_{ct}(n)$	Cosine transform coefficient for x_i
X_n	Transformed value of x_i
$X_{m,n}$	Transformed value of x_{ij}
$X_s(k)$	SAL transform coefficient for x_i
$X_{st}(k)$	Sine transform coefficient for x_i
$Z(\tau)$	Logical convolution
d, Δ	Increment
β	Angle
ϵ	Dielectric constant
η	Relative bandwidth
θ	Angle; normalised time (t/T)
λ	Wavelength
μ	Magnetic permeability
π	3.14159
σ	Conductivity
τ	Time delay
$\tau(\omega)$	Frequency transfer function
ϕ	Angle
ω	$2\pi f$, angular frequency
$\circledast$	Dyadic convolution operator
$\oplus$	Modulo-2 addition
$\otimes$	Kronecker product
$\leftrightarrow$	Transform operator

Part One

Theory and Practice

Chapter 1

The Sequency Functions

1.1 Introduction

In our attempts to understand natural events and to communicate information about them, we find it necessary to carry out analysis, classification and comparison. Since much of this information is acquired in the form of signals which are functions of time, a body of signal theory has been developed to consider these data in terms of time and frequency. The orthogonal system of sine and cosine functions plays an important role in this theory, and Fourier analysis is probably the most important mathematical tool.

In recent years, aided by the power and capability of digital computation, other orthogonal systems of functions have found a place in this process of understanding and communication and other methods of analysis discovered. Our approach to a problem in analysis can now be much broader and we can consider, in practical terms, the application of a whole series of orthogonal functions and transformations, each of which has its own particular relevance.

One particular area of interest lies in the representation of a signal (which can be continuous in either time or space) through the superposition of members of a set of simple functions which are easy to generate and define. It is important, for example, to be able to represent waveforms used for communication in this way since generalisations can then be made from the values of the set of functions required, thus permitting equipment design characteristics to be evaluated. In the following we consider the characteristics of the various function sets we could choose for this purpose, commencing with a consideration of their orthogonality.

1.2 Orthogonality

Only *orthogonal* sets of functions can be made to synthesise completely any time function to a required degree of accuracy. Further, the characteristics of an orthogonal set are such that identification of a particular member of the set contained in a given time function can be made by using quite simple mathematical operations on the function.

We can consider a signal time function $x(t)$ defined over a time interval $(0, T)$ as being represented by an orthogonal series $S_n(t)$. Thus

$$x(t) = \sum_{t=0}^{T} C_n S_n(t) \tag{1.1}$$

where C_n is a number indicative of the magnitude of the series constituents.

The series $S_n(t)$ ($n = 0, 1, 2, \ldots$) is said to be orthogonal with weight K over the interval $0 \geqslant t \geqslant T$ if

$$\int_0^T K \cdot S_n(t) S_m(t) \, dt = \begin{cases} K & \text{if } n = m \\ 0 & \text{if } n \neq m \end{cases} \tag{1.2}$$

when n and m have integer values and K is a non-negative constant (or fixed function) which does not depend on the indices m and n. If the constant K is one, then the set is normalised and called an *orthonormal* set of functions. A non-normalised set can always be converted into an orthonormal set.

Since only a finite number of terms N is possible for a practical realisation of the series given by Eq. (1.1), it is necessary to choose the coefficients C_n in order to minimise the mean-square approximation error:

$$\text{MSE} = \int_0^T [x(t) - \sum_{n=0}^{N-1} C_n S_n(t)]^2 \, dt \tag{1.3}$$

which is realised by making

$$C_n = \frac{1}{T} \int_0^T x(t) S_n(t) \, dt \tag{1.4}$$

It is desirable that this error monotonically decreases to zero as N becomes very large. This is the case for a <u>complete</u> orthogonal function series.

A complete orthonormal function series $S_n(t)$ is always a <u>closed</u> series. That is, there exists no quadratically integrable function $x(t)$ where

$$0 < \int_b^a x^2(t) \, dt < \infty \tag{1.5}$$

and for which the equality

$$\int_b^a x(t), S_n(t) \, dt = 0 \tag{1.6}$$

is satisfied for all values of n.

More generally, a function series, whether normal and orthogonal or not, is said to be _complete_ or _closed_ if no function exists which is orthogonal to every other function of the series unless the integral of the square of the function is itself zero [1]. We shall be considering the completeness of a function series again later in connection with specific series.

Incomplete orthogonal function series do not converge and therefore cannot represent exactly any given time function, although they may have other properties of equal importance. For example, the output of a low-frequency filter can comprise an incomplete orthogonal series of $(\sin x)/x$ functions. Another example of an incomplete series is one comprised of Rademacher functions. These are a set of simple rectangular functions which, as we shall see later, play an important part in the generation of other function series.

By using an orthogonal time series representation, the signal can be expressed as a limited set of coefficients or spectral numbers. Quite considerable reductions can be made in the number of coefficients needed for complete representation in this way without losing the identity of the signal. It is also possible to use the orthogonal property of a data series to identify specific components of the series.

In summary we may state that any set of functions which is capable of being integrated and of integrable square modulus may be used to form a closed set of linear combinations of the functions which will be normal and orthogonal. Also these combinations will be found to be obtainable from a limited number of constituent functions. The circular functions are the most well known of these, and their orthogonality will be considered in the next section.

1.2.1 Sine–cosine functions

Let us consider a series of sine–cosine functions, the first six numbers of which are shown in Fig. 1.1. The significance of orthogonality may be seen if we take the products of pairs of such functions over a limited time interval $0 \leqslant t \leqslant T$. Thus if we let

$$S_n(t) = \sqrt{2} \cos 2\pi nt \quad \text{or} \quad \sqrt{2} \sin 2\pi nt$$
$$S_m(t) = \sqrt{2} \cos 2\pi mt \quad \text{or} \quad \sqrt{2} \sin 2\pi mt \quad (1.7)$$

then from Eq. (1.2)

$$\int 2 \cos 2\pi mt \cos 2\pi nt \, dt$$
$$= \int (\cos(m+n)2\pi t + \cos(m-n)2\pi t) \, dt = 0 \quad (1.8)$$

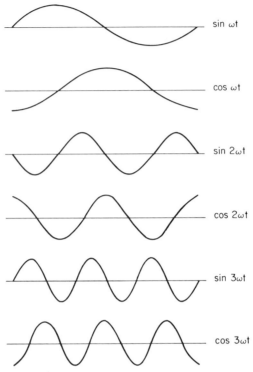

FIG. 1.1. A set of sine–cosine functions.

if $m \neq n$ and both m and n are integers, since the average value of a cosinusoidal waveform over an integer number of periods is zero.

Similarly,

$$\left. \begin{array}{l} \int 2 \sin 2\pi mt \sin 2\pi nt \, dt \\ \int 2 \sin 2\pi mt \cos 2\pi nt \, dt \\ \int 2 \cos 2\pi mt \sin 2\pi nt \, dt \end{array} \right\} \text{also equal } 0 \quad \text{if} \quad m \neq n \quad (1.9)$$

However, for $m = n$ then

$$\int 2 \sin^2 2\pi nt \, dt = \int 2 \cos^2 2\pi nt \, dt = 2$$

over the interval $(0, T)$.

1.2 Orthogonality

Since any time series can be expressed as a summation of sinusoidal components (Fourier series), viz.,

$$x(t) = \frac{a_0}{2} + \sum_{k=1}^{\infty} \{a_k \cos(k\omega_0 t) + b_k \sin(k\omega_0 t)\} \quad (1.10)$$

then multiplication by $\cos k\omega_0 t$ or $\sin k\omega_0 t$ and integrating over the period $2\pi/\omega_0$ will enable the Fourier coefficients a_k and b_k to be extracted.

This follows from the above, where it is seen that the process of integration will reduce all the other sine–cosine products to zero. The orthogonal feature of sine–cosine representation, therefore, solves the problem of identifying a particular sinusoidal component from a composite waveform comprising the summation of many elements and hence is the key to spectral decomposition, filtering and similar operations.

1.2.2 Incomplete function sets

A function set is <u>complete</u> if the mean-square error of the signal representation $x(t)$ converges to zero with increasing number of terms, viz.,

$$\lim_{N \to \infty} \int_0^T [x(t) - \sum_{n=0}^{N=1} C_n S_n(t)]^2 \, dt = 0 \quad (1.11)$$

The *mean-square error*

$$\text{MSE} = \int_0^T [x(t) - \sum_{n=0}^{N=1} C_n S_n(t)]^2 \, dt \quad (1.12)$$

depends on the system of functions chosen for the linear approximation to the data series or waveform. If the shape of this waveform is similar to that of the functions used, then the MSE will be small.

An alternative definition for completeness of a function set is that of *Parseval's theorem* or the completeness theorem, which will be given later. A physical meaning for this theorem, applicable to a complete series, is to state that for this case the energy contained within the series will be the same whether expressed in the time or transformed domain.

One example of an incomplete series is the set of orthogonal block pulses shown in Fig. 1.2. The conditions of Eq. (1.2) obviously obtains here since only one of the signals is allowed to differ from zero at any one time, although the system is not complete.

The definitions of <u>completeness</u>, normality and orthogonality may be applied to all functions over a semi-infinite interval $(0, \infty)$ or the complete infinite interval $(-\infty, +\infty)$ as well as to functions defined over a finite interval $(+T/2, -T/2)$ or $(0, T)$. The circular functions can apply to any of these

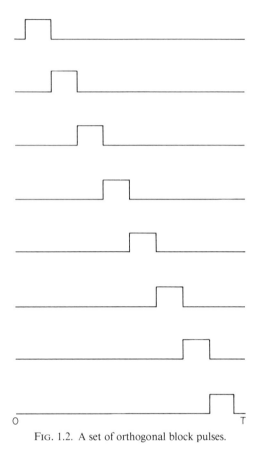

FIG. 1.2. A set of orthogonal block pulses.

cases, but the Walsh, Haar and certain other functions are limited to finite intervals only.

The limitation in the case of Walsh and other functions confers the advantage that a time-limited signal composed of a limited number of orthogonal functions will occupy a finite section of the transformed domain. With the circular functions, on the other hand, a finite time signal occupies an infinite frequency function in the transformed domain. This has a relevance in signal reconstruction for a given accuracy and in the analysis of nonstationary waveforms.

1.2.3 Rademacher functions

An important, incomplete but orthogonal function set are the Rademacher functions [2]. These represent a series of rectangular pulses or square waves having unit mark–space ratio. The first six of these is shown in Fig. 1.3. The first function RAD(0, t) is equal to one for the entire interval $0 \leq t \leq T$. The next and subsequent functions are square waves having odd symmetry. The incompleteness of the series can be demonstrated if we consider the summation of a number of Rademacher functions. This composite waveform will also have odd symmetry about the centre and similar, even symmetry functions required for completeness cannot be developed.

Rademacher functions have two arguments n and t such that RAD(n, t) has 2^{n-1} periods of square wave over a normalised time base $0 \leq t \leq 1$. The amplitudes of the functions are $+1$ and -1. They can be derived from sinusoidal functions which have identical zero crossing positions. Thus

$$\text{RAD}(n, t) = \text{sign}[\sin(2^n \pi t)] \qquad (1.13)$$

and may be obtained from a sinusoidal waveform of appropriate frequency by amplification followed by hard limiting. They are important principally because other complete series, such as the Walsh series, can be derived from them.

1.2.4 Walsh functions

The Walsh functions [3] form an ordered set of rectangular waveforms taking only two amplitude values $+1$ and -1 and are another example of an

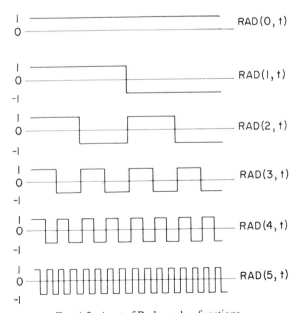

FIG. 1.3. A set of Rademacher functions.

orthonormal set of functions. Unlike the Rademacher functions, the Walsh rectangular waveforms do not have unit mark–space ratio. They are defined over a limited time interval T known as the *time base*, which requires to be known if quantitative values are to be assigned to a function. Like the sine–cosine functions, two arguments are required for complete definition, a time period t (usually normalised to the time base as t/T) and an ordering number n related to frequency in a way which is described later. The function is written

$$\text{WAL}(n, t), \quad n = 0, 1, \ldots, N - 1 \qquad (1.14)$$

and for most purposes a set of such functions is ordered in ascending value of the number of zero crossings found within the time base. Figure 1.4 shows the first eight of these with the ordering arranged in this way.

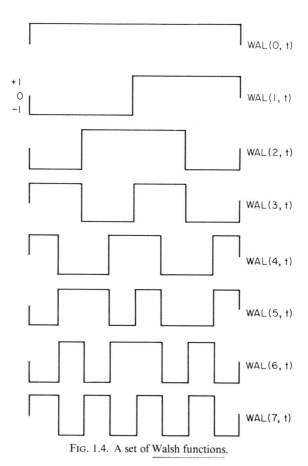

FIG. 1.4. A set of Walsh functions.

1.2 Orthogonality

The orthogonality of the discrete Walsh series can be proved as follows:

First an expression for the discrete Walsh function having $N = 2^p$ terms will be stated in terms of a continued product [4] as

$$wal(n,t) = \text{WAL}(n_{p-1}, n_{p-2}, \ldots, n_0; t_{p-1}, t_{p-2}, \ldots, t_0)$$

$$= \prod_{r=0}^{p-1} (-1)^{n_{p-1-r}(t_r + t_{r+1})} \qquad 0 \le n < N \qquad (1.15)$$
$$0 \le t < N$$

where n and t are the arguments of the function expressed in binary notation.

The sum of the products of any two discrete Walsh functions is given as the binary summation

$$\sum_{t_{p-1}}^{1} \sum_{t_{p-2}}^{1} \cdots \sum_{t_0=0}^{1} \text{WAL}(m_{p-1}, m_{p-2}, \ldots, m_0; t_{p-1}, t_{p-2}, \ldots, t_0)$$

$$\times \text{WAL}(n_{p-1}, n_{p-2}, \ldots, n_0; t_{p-1}, t_{p-2}, \ldots, t_0) \qquad (1.16)$$

Substituting Eq. (1.15) in (1.16) gives for the binary product–sum

$$\sum_{t_{p-1}}^{1} \sum_{t_{p-2}}^{1} \cdots \sum_{t_0=0}^{1} \prod_{r=0}^{p-1} (-1)^{(n_{p-1-r} + m_{p-1-r})(t_r + t_{r+1})}$$

$$= \prod_{r=0}^{p-1} \sum_{t_r=0}^{1} (-1)^{(n_{p-1-r} + m_{p-1-r})(t_r + t_{r+1})}$$

$$= \prod_{r=0}^{p-1} \{1 + (-1)^{(n_{p-1-r} + m_{p-1-r})}\} \qquad (1.17)$$

Now if each $n_t = m_t$ and it is remembered that only two values are possible, zero or one, then Eq. (1.17) becomes

$$\prod_{r=0}^{p-1} (1 + 1) = 2^p = N$$

If at least one $n_t \ne m_t$, then at least one term in the product given by Eq. (1.16) is zero, giving a zero product. In terms of decimal indices we have for the product of two Walsh terms

$$\sum_{t=0}^{N-1} \text{WAL}(m, t)\, \text{WAL}(n, t) = \begin{cases} N & \text{for } n = m \\ 0 & \text{for } n \ne m \end{cases} \qquad (1.18)$$

Hence the Walsh functions can be seen to form an orthogonal set which can be normalised by division by N to form an orthogonal system.

1.2.5 Haar functions

Haar functions also form a complete orthonormal function set of rectangular waveforms and were proposed originally by Haar [5] in 1910. The functions have several important properties, including the ability to represent a given function with few constituent terms to a high degree of accuracy.

They have three possible states 0, $+A$ and $-A$, where $\pm A$ is a function of $\sqrt{2}$. Thus, unlike the Walsh functions, the amplitude of the functions varies with their place in the series.

Haar functions may be presented over the interval $0 \leq t \leq 1$ as

$$\text{HAR}(n, t) \qquad (1.19)$$

where n also identifies the function in terms of zero crossings. For complete definition it is also necessary to give some information concerning the amplitude of the function since this is no longer confined to $+1$ and -1 as with the Walsh function series. An alternative representation giving zero crossing order equality with the Walsh series is given later.

The first eight Haar functions are shown in Fig. 1.5. The functions are

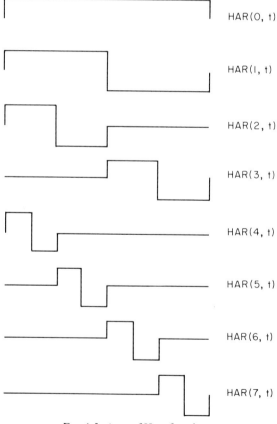

FIG. 1.5. A set of Haar functions.

1.2 Orthogonality

orthogonal and orthonormal and obey the condition for orthogonality

$$\int_0^1 \mathrm{HAR}(m, t)\, \mathrm{HAR}(n, t)\, dt = \begin{cases} 1 & \text{for} \quad n = m \\ 0 & \text{for} \quad n \neq m \end{cases} \quad (1.20)$$

1.2.6 Slant functions

A special function series known as the *slant function* series was introduced by Enomoto and Shibata [6] for image transmission purposes. The functions have a finite but large number of possible states, as may be seen from the series for $N = 8$ shown in Fig. 1.6, and are similar to the general form of the Walsh-ordered series.

They may be represented over the interval

$$0 \leq t \leq 1 \quad \text{as} \quad \mathrm{SLA}(n, t) \quad (1.21)$$

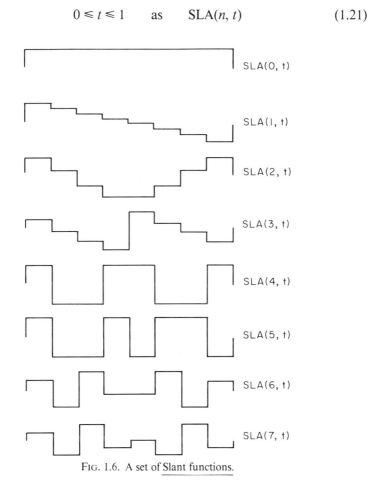

FIG. 1.6. A set of Slant functions.

The rationale for this function series lies in its transform characteristics which permit a compaction of the image energy to only a few transformed samples, thus improving the efficiency of image data transmission in this form. It is recognised that a high degree of energy compactness will result if the elements of a function series resemble the brightness characteristics of the typical horizontal or vertical lines of a scanned image. Two common characteristics are lines which have a nearly constant gray level over much of their length and those for which the image brightness increases or decreases linearly over the length of a line. Whilst the Walsh functions can match the constant gray characteristics, other functions needed to be added to the series to provide a discrete sawtooth representation for the linearly changing gray-level characteristic. Although details of the slant orthogonal function are given later, it should be noted that subsequent developments in fast cosine and sine transformations for image representation indicate a less active role for future applications of the slant transform (see Subsection 6.2.1).

1.2.7 Mixed and other functions

There are some advantages in deriving orthogonal function series from a combination of Haar, Walsh or other functions in much the same way as the slant series is obtained from the admixture of a sequency-ordered Walsh series and a stepped sawtooth waveform. Some examples of these mixed function series will be given later.

An orthogonal function set which has ideal mean-square error characteristics for signal representation is the Karhunen–Loève series [7]. It provides a generally optimum transform since it compacts the most energy into a few coefficients, as the slant transform is designed to do for the special case of image transformation. Unfortunately, implementation of the transform involves the determination of eigenvalues and corresponding eigenvectors of the covariance matrix, which allows for no simple calculation algorithm based on redundancy in the repeated calculations, the basis of fast transform calculations. The efficiency of signal representation for the Karhunen–Loève function series is, however, recognised as a useful criteria by which the value of other orthogonal series can be measured. Consequently, the efficiency of a given series is often compared in this way.

Finally, there are a number of polynomial functions which can be made orthogonal by multiplication by a weighting factor. These orthogonal polynomials consist of a series, $f_n(x)$ $(n = 0, 1, 2, . . .)$, where n is the degree of the polynomial. This class contains many special functions commonly encountered in practical applications, e.g., Chebychev, Hermite, Laguerre, Jacobi, Gegenbauer and Legendre polynomials [8, 9]. None of these contains the essential simplicity of the sequence functions described earlier where members of each of these classes assume a single value having either a positive or a negative sign which has the effect of reducing multiplicative operations on the series to an appropriate sequence of sign changes. This simplicity gives rise to favourable repercussions in calculation and in digital hardware, which will be referred to time and again in the following pages.

1.3 The Walsh function series

In this and subsequent sections more detailed consideration will be given to the Walsh and other non-sinusoidal orthogonal functions preparatory to discussing their domain transformation in Chapter 2.

Earlier we identified these functions by means of a common feature, namely, their orthogonality. A second characteristic that they share lies in the parameter which distinguishes between individual functions in a given series through the number of times the function passes through zero over the time base. This is known as *sequency*, and for convenience the non-sinusoidal functions discussed here will be referred to as sequency functions.

Sequency is a term proposed originally by Harmuth [10] to describe a periodic repetition rate which is independent of waveform. It is defined as 'one-half of the average number of zero crossings per unit time interval'. From this we see that frequency can be regarded as a special measure of sequency applicable to sinusoidal waveforms only.

If the function terms are ordered in ascending value of sequency, then this is known as *sequency ordering* and has similarities to the familiar sine–cosine series (compare the order of the Walsh functions given in Fig. 1.4 with the set of sinusoidal functions of Fig. 1.1).

1.3.1 Definitions

A notation for the Walsh series was given in Subsection 1.2.4. An alternative notation has been introduced by Harmuth [10] to classify the Walsh functions in terms of even and odd waveform symmetry, viz.,

$$\text{WAL}(2k, t) = \text{CAL}(k, t), \quad \text{WAL}(2k - 1, t) = \text{SAL}(k, t)$$

$$k = 1, 2, \ldots, N/2 \quad (1.22)$$

which defines two further Walsh series having close similarities with the cosine and sine series. Figure 1.7 gives the first 32 of the Walsh function series, arranged in sequency order and with the alternative CAL and SAL terms shown.

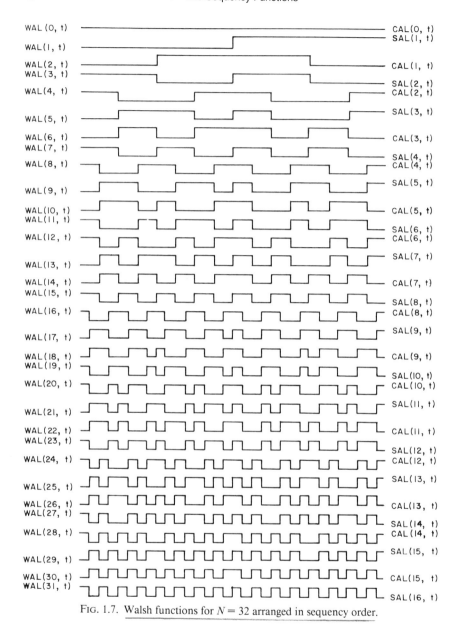

FIG. 1.7. Walsh functions for $N = 32$ arranged in sequency order.

As indicated in this diagram, the normalised Walsh functions are symmetrical about the centre. When the range of the function is defined as $-\tfrac{1}{2} \leq t \leq \tfrac{1}{2}$, the functions are either <u>directly symmetrical</u> (CAL functions) or <u>inversely symmetrical</u> (SAL functions). In this latter case, the ones found in

1.3 The Walsh Function Series

the left-hand side are mirrored by zeros in the right-hand side and vice versa. This enables a *symmetry relationship* to be stated as

$$\text{WAL}(n, t) = \text{WAL}(t, n) \tag{1.23}$$

As will be discussed later, the practical importance of this is that the transform and its inverse represent the same mathematical operation and thus simplify the derivation and application of the transform.

We also see from Fig. 1.7 that the number of zero crossings in the interval $-\frac{1}{2} \leq t \leq \frac{1}{2}$ is $2k$ so that k represents sequency in CAL and SAL ordering.

The similarities between the circular and Walsh functions are also seen in the expressions for the function series. Using the Fourier series expansion, we can express a time series $x(t)$ as the sum of a series of sine–cosine functions each multiplied by a coefficient giving the value of the function for that series, viz.,

$$x(t) = \frac{a_0}{2} + \sum_{k=1}^{\infty} (a_k \cos(k\omega_0 t) + b_k \sin(k\omega_0 t)) \tag{1.24}$$

where

$$\frac{a_0}{2} = \frac{1}{T} \int_0^T x(t) \, dt$$

$$a_k = \frac{2}{T} \int_0^T x(t) \cos k\omega_0 t \, dt \tag{1.25}$$

$$b_k = \frac{2}{T} \int_0^T x(t) \sin k\omega_0 t \, dt$$

The coefficients a_k and b_k represent the peak amplitudes of the spectral components of $x(t)$. A set of these coefficients can form a further series $x(k)$, which expresses $x(t)$ in the frequency domain.

We can also express a time series $x(t)$ in a similar way in terms of the sum of a series of Walsh functions, viz.,

$$x(t) = a_0 \, \text{WAL}(0, t) + \sum_{n=1}^{N-1} a_n \, \text{WAL}(n, t) \tag{1.26}$$

where

$$\frac{a_0}{2} = \frac{1}{T} \int_0^T x(t) \, \text{WAL}(0, t) \, dt$$

$$a_n = \frac{1}{T} \int_0^T x(t) \, \text{WAL}(n, t) \, dt \tag{1.27}$$

or from Eq. (1.22) by using the sum of two series for CAL and SAL terms having $N/2 - 1$ and $N/2$ values, respectively,

$$b_j = \frac{1}{T} \int_0^T x(t) \, \text{CAL}(j, t) \, dt$$

and

$$a_i = \frac{1}{T} \int_0^T x(t) \, \text{SAL}(i, t) \, dt$$

(1.28)

Using the SAL and CAL forms, we can obtain an expression for the Walsh series similar to that given for the sine–cosine series in Eq. (1.24), viz.,

$$x(t) = a_0 \, \text{WAL}(0, t) + \sum_{i=1}^{N/2} \sum_{j=1}^{N/2-1} (a_i \, \text{SAL}(i, t) + b_j \, \text{CAL}(j, t)) \quad (1.29)$$

The two new series of a_i and b_j coefficients taken together express $x(t)$ in the sequency domain. Note that $\text{WAL}(0, t) = \text{CAL}(0, t)$ so that, in the expression, there is one less CAL term than there are SAL terms in the summation.

The derivation of these coefficient series is referred to as <u>decomposition into the spectral components of $x(t)$</u>, although these components are now no longer sinusoidal in form.

We may note that although it is possible to combine the sine and cosine elements into a single complex variable, $\exp(\pm jk\omega_0 t)$ (where $j = \sqrt{-1}$), expressing the same frequency, this is not possible with Walsh functions due to the <u>absence of a shift theorem similar to that found in circular function theory</u>. Consequently, the two separate series, which are developed from the SAL and CAL functions of the Walsh series, are needed to express fully the sequency behaviour of $x(t)$.

Synthesis of a complex waveform using the principle of superposition from a linear set of functions is obtained by using the Walsh series in a manner analogous to Fourier synthesis. A stepped equivalent waveform is obtained which approximates the original waveform more closely as the number of superimposed series is increased. As an example Fig. 1.8 shows a very simple approximation of a sinusoidal waveform from the three principal Walsh series, each having an appropriate amplitude. The number of terms required for a given MSE is dependent on the characteristics of the reconstructed waveform in relation to those of the constituent series. Consideration is given to this in <u>Subsection 3.5.2</u>, where <u>equivalent Fourier and Walsh synthesis</u> are discussed.

1.3.2 Function ordering

As we shall see later the generation of a Walsh function series can be carried out in a number of ways and the order of the functions produced can be different to the sequency order shown in Fig. 1.7.

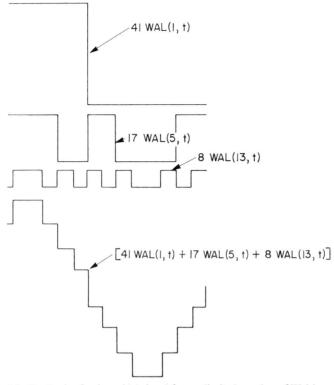

FIG. 1.8. Synthesis of a sinusoidal signal from a limited number of Walsh series.

Three main ordering conventions are in common use, and others have been proposed and employed for special purposes. Unless we are clear about the convention used, confusion can arise when results and algorithms from different sources are compared. Whilst the sequency order shown in Fig. 1.7 is closest to our practical experience with other orthogonal functions (e.g., sinusoidal functions), it does not offer the most convenient order for analytical and computational purposes. All three forms of ordering are used, and we need now to study these and their relationships with each other. The three forms are the following.

(a) *Sequency order* (ordered form, Walsh order, Walsh–Kaczmarz order) This was Walsh's original order for his function WAL(n, t), and he arranged the components in ascending order of zero crossings (Fig. 1.7). It is directly related to frequency where we find that Fourier components are also arranged in increasing harmonic number (zero crossings divided by two). The advantage of this order is that derivation of the alternate CAL and SAL functions shown resembles that of orthonormal cosine and sine functions in Fourier analysis and hence permits suitable comparisons to be made.

(b) *Dyadic order* (Paley order) This is the order obtained by generation from successive Rademacher functions. It was first used by Paley [11] and will be referred to as the Paley-ordered function

$$\text{PAL}(n, t) \tag{1.30}$$

Dyadic ordering has certain analytical and computational advantages noted by Fine, Gibbs, Ahmed, Yuen and others and is used for most mathematical discussions [12]. In particular, Gibbs has shown that the Paley-ordered functions may be defined as the eigenfunctions of a logical differential operator and that this definition is of value in the mathematical development of the theory [13].

(c) *Natural order* (Hadamard order) This ordering follows the Hadamard matrix derived from successive Kronecker products (of more later) and was originally proposed by Henderson [14]. It is the ordering that is obtained if one computes fast Walsh transforms without sorting in the manner of the Cooley–Tukey fast Fourier transform algorithm; hence it is computationally advantageous. However, as we shall see later, it can be defined only for matrices of a known size since a bit-reversal operation is required to relate to Walsh or dyadic ordering and this is necessarily dependent on the number of digits involved.

The natural order is referred to here as the function

$$\text{HAD}(n, t) \tag{1.31}$$

The relationships between these various orders is considered in Subsection 1.3.5.

Dyadic and natural ordering are used in theoretical mathematical work, in image transmission and for computational efficiency. Sequency order is favoured for communications and signal-processing work such as spectral analysis and filtering.

Figures 1.9 and 1.10 show the first eight Walsh functions arranged in dyadic and natural order, respectively. These may be compared with the sequency order shown in Fig. 1.7, in which quite considerable changes in relative positions for the various functions are seen.

In the derivation of the Walsh series and for certain special requirements other orderings may be encountered. Some are related simply to *bit-reversal* of one of the ordering forms discussed earlier. Thus, if we consider the binary equivalents to an ascending series 0, 1, 2, 3, 4, 5 as 000, 001, 010, 011, 100,

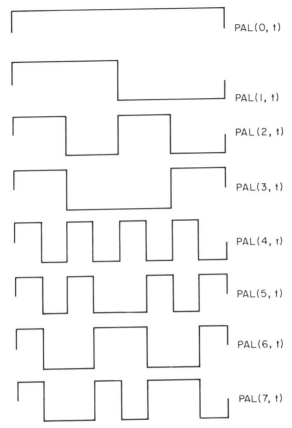

FIG. 1.9. Walsh functions for $N = 8$ arranged in dyadic order.

101, then reversing the order of the binary digits gives 000, 100, 010, 110, 001, 101, which are expressed in decimal order as 0, 4, 2, 6, 1, 5, respectively. This is the bit-reversed order for the first six integer values. Rearrangement into the required order for a given set N is made through suitable routing in the case of hardware derivation or a bit-reversal software routine which may precede or follow the transform routine.

Recent developments in Walsh matrix theory have shown that the Walsh functions can be arranged in several orderings beyond the three common groupings discussed [15]. One important member of this group which has the advantage of providing discrimination between signals having even or odd symmetry is known as CAL – SAL ordering [16]. Here the first half of the rows of the function series is arranged with increasing number of even zero

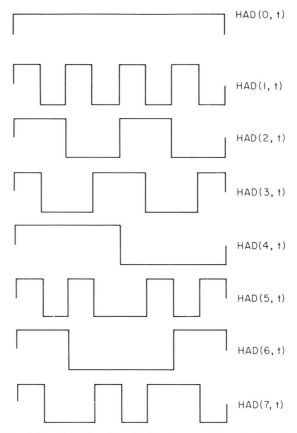

FIG. 1.10. Walsh functions for $N = 8$ arranged in natural order.

crossings (corresponding to CAL functions) whilst the second half has a decreasing number of odd zero crossings (corresponding to SAL functions). This is shown in Fig. 1.11 for $N = 8$.

The relationship with the sequency-ordered Walsh functions is given as

$$\begin{aligned} \text{WAL}_{cs}(n, t) &= \text{WAL}(2n, t) \\ \text{WAL}_{cs}(N/2 + n, t) &= \text{WAL}(N - 2n - 1, t) \end{aligned} \quad (1.32)$$

where $n = 0, 1, \ldots, N/2 - 1$.

A further ordering arrangement which is of value in digital logic design [17] is known as Rademacher–Walsh ordering and is shown in Fig. 1.12, again for $N = 2^p = 8$. The first $p + 1$ functions consist of Rademacher functions $\text{RAD}(0, t)$, $\text{RAD}(1, t)$, $\ldots$, $\text{RAD}(p, t)$ arranged in ascending se-

1.3 The Walsh Function Series

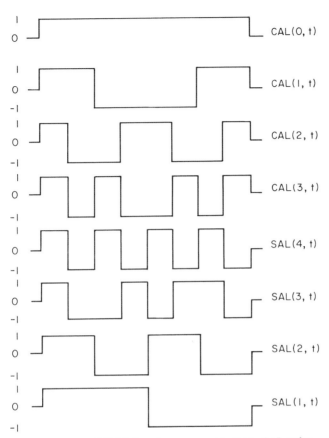

FIG. 1.11. A set of Walsh functions arranged in CAL–SAL order.

quency order and are known as the primary set of functions. Subsequent functions, known as the secondary set, may be formed from combinations of the primary set in a way which will be described later (Section 8.2).

1.3.3 Phase of the ordered set

The diagram in Fig. 1.7 is arranged to emphasise the phase similarity with an ordered set of sine–cosine functions and will be referred to as *Harmuth phasing*. However, if the series is derived directly from the Hadamard matrix or from Rademacher products, as will be described later, then the functions will be phased such that they all start at a $+ 1$ level, and this will be referred to as *positive phasing*. Figures 1.9 and 1.10 have been presented in this way.

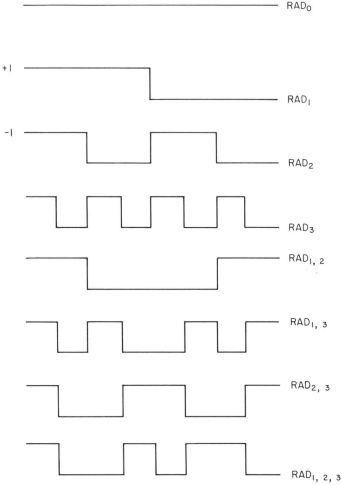

Fig. 1.12. A set of Rademacher–Walsh functions.

1.3.4 Derivation of Walsh series

The function series can be obtained in several different ways, each of which has its own particular advantages. The methods considered here are those

    (i)   from recursive relations [10, 20],
    (ii) from products of Rademacher functions [29],
    (iii) through the Hadamard matrix [18], and
    (iv) by the use of Boolean synthesis [19].

The difference method gives the function directly in sequency order. The other methods give results in dyadic order or natural order. A method of converting these to sequency order will be described later.

All these derivations are, of course, mathematical processes for which computational algorithms can be developed and the series produced by using the digital computer or obtained directly by using digital logic. However, owing to the simplicity of the fast Walsh transform algorithm and the speed with which this can be implemented on the digital machine, the generation of a function or series of functions can also be obtained simply by transforming a unit input sample at the appropriate position in the input vector by using the fast Walsh transform. This will become apparent when signal flow diagrams for this algorithm are studied.

(i) *From recursive relations* In Harmuth's derivation [10] it is assumed that the normalised time base is used referenced to its centre, i.e., $-\frac{1}{2} \leq \theta \leq \frac{1}{2}$. A given Walsh function is defined from its preceding harmonic function so that, commencing with a definition of WAL$(0, \theta) = 1$ within the time base and 0 outside the time base, the entire set of Walsh functions can be obtained by an iterative process.

The difference equation is given as

$$\text{WAL}(2j + q, \theta) = (-1)^{(j'/2+q)}$$
$$\times [\text{WAL}(j, 2(\theta + \tfrac{1}{4})) + (-1)^{(j+q)} \text{WAL}(j, 2(\theta - \tfrac{1}{4}))] \quad (1.33)$$

where $j = 0, 1, 2, \ldots$ and $q = 0$ or 1. Here $j'/2$ means the largest integer smaller or equal to $j/2$. Thus for $j = 1$, $j'/2 = 0$.

A somewhat easier notation is obtained if we reference time to the commencement of the generated function to avoid negative values of θ. Thus, if we replace θ by t and define this as $0 \leq t \leq 1$, then, as before, WAL$(0, t) = 1$ within this time base and 0 outside, so that the difference equation can be restated as

$$\text{WAL}(2j + q, t) = (-1)^{(j'/2+q)}$$
$$\times [\text{WAL}(j, 2t) + (-1)^{(j+q)} \text{WAL}(j, 2(t - \tfrac{1}{2}))] \quad (1.34)$$

which for N equally spaced discrete points (where $N = 2^p$ and $n = 0, 1, 2, \ldots, N-1$) can be written

$$\text{WAL}(2j + q, n) = (-1)^{(j'/2+q)}$$
$$\times [\text{WAL}(j2n) + (-1)^{(j+q)} \text{WAL}(j, 2(n - N/2))] \quad (1.35)$$

Commencing with the known WAL$(0, n) = 1$ within the time base (i.e., $j = 0, q = 1$) for $n \leq N/2$, its value for WAL$(j, 2n)$ will be 1 and for $n > N/2$

the function falls outside the time base and will be 0. Similarly with WAL(j, $2(n - N/2)$) for $n \leq N/2$ the function again falls outside the time base and will become 0, whilst for $n > N/2$ the value is 1. The sign of these functions will be modified by the factors $(-1)^{(j'/2+q)}$ and $(-1)^{(j+q)}$ in accordance with Eq. (1.35).

This procedure is summarised to give the required function WAL(1, n) in Table 1.1. Note that values of $n < 0$ or $n > 1$ must result in any Walsh function having a value of 0. From this result WAL(2, n) can be obtained in a similar manner and the process repeated to obtain further functions in the series. The operation of this difference equation may be considered as equivalent to compressing the previous Walsh function WAL(p, $2n$) into the left-hand part of the time base by selecting alternate points and after left adjustment by adding to these on the right-hand side a similar valued set of points but all having an opposite sign.

An alternative recursive derivation is due to Chien [20]. It enables an explicit representation of a Walsh function of any order to be obtained without first having two construct Walsh functions of a lower order. Let WAL(n, t) be represented as

$$\text{WAL}(n, t) = (-1)^{(2^k t)+(2^j t)+(2^i t)+(2^h t)+\cdots} \quad (1.36)$$

$k > j > i > h > \cdots$ The parameters $k, j, i, h, \ldots$ are determined successively as follows.

First k is obtained so that $2^k > n > 2^{k-1}$; from this j is defined so that $2^k - 2^{j-1} > n \geq 2^k - 2^j$ and i so that $2^k - 2^j + 2^i \geq n > 2^k - 2^j + 2^{i-1}$ and h so that $2^k - 2^j + 2^i - 2^{h-1} > n \geq 2^k - 2^j + 2^i - 2^h$ and so on. The process

TABLE 1.1. Derivation of the Walsh function by the difference method.

n	WAL(j, $2n$) = A	$(-1)^{(j+q)} = B$	WAL(j, $2(n-N/2)$) = C	$(-1)^{[j/2]+q} = D$	WAL(1, n) = $D(A+BC)$
0	1	-1	0	-1	-1
1	1	-1	0	-1	-1
2	1	-1	0	-1	-1
3	1	-1	0	-1	-1
$\vdots$	$\vdots$	$\vdots$	$\vdots$	$\vdots$	$\vdots$
$N/2-1$	1	-1	0	-1	-1
$N/2$	0	-1	1	-1	1
$\vdots$	$\vdots$	$\vdots$	$\vdots$	$\vdots$	$\vdots$
$N-1$	0	-1	1	-1	1

stops when the alternative sum has converged to n (for even values of n) or $n + 1$ (for odd values). It will be seen that since the factor $(-1)^{2kt}$ can be considered as an explicit representation of a Rademacher function of order $k - 1$, this method is essentially the same as the product of Rademacher functions, which we shall consider next.

(ii) *From products of Rademacher functions* Rademacher functions were introduced in Subsection 1.2.3 and are illustrated in Fig. 1.3. Although they form an incomplete series having odd symmetry, it is possible to form functions from them which will exhibit either odd or even symmetry. Hence a complete series can be developed from the incomplete set of functions. In particular a complete set of Walsh functions in dyadic order can be obtained from selected Rademacher function products as follows.

The product series for the Rademacher functions is expressed as

$$\text{PAL}(n, t) = \prod_{i=1}^{m} \text{RAD}(i, t)^{b_i} \tag{1.37}$$

when n is expressed as a binary number

$$n = b_m 2^m + b_{m-1} 2^{m-1} + \cdots b_1 2^1 + b_0 2^0 \tag{1.38}$$

and $b_i = 0$ or 1.

Thus, to find PAL(13, t), we can write

$$\text{PAL}(13, t) = \text{RAD}(4, t)\, \text{RAD}(3, t)\, \text{RAD}(1, t)$$

since binary 13 is 1101 and the 4, 3 and 1 refer to ones found in the binary bit positions.

This derivation will give a dyadic-ordered series having positive phasing. If the open interval for the set of Rademacher functions is defined as $-\frac{1}{2} \leq t \leq \frac{1}{2}$, then the functions will have the inverse sign to that shown in Fig. 1.3. Their products will then give rise to the Harmuth phasing for the dyadic-ordered series.

The products actually refer to Rademacher functions represented as a string of -1 and $+1$'s. If we adopt the convention that -1 equals binary 0 and $+1$ equals binary 1, then the products become sums, and we write

$$\text{PAL}(n, t) = \sum_{i=1}^{m} b_i\, \text{RAD}(i, t) \tag{1.39}$$

But the summations for the Rademacher functions are expressed as *modulo-2 addition*, i.e., binary sums without carry, and obey the rules

$$0 \oplus 0 = 0, \quad 0 \oplus 1 = 1, \quad 1 \oplus 0 = 1 \quad \text{and} \quad 1 \oplus 1 = 0$$

A recursive method of generating a sequency-ordered set from Rademacher functions is described by Harmuth [10]. A simple definition is given by Lackey and Meltzer [21] to enable any given Walsh function to be obtained from a given Rademacher function product. They note that the sequency order (n) is related to the natural order by means of the *Gray code,* given in Table 1.2. To find the value of bit position (i) of the sequency number n expressed in the Gray code, we need to add bit (i) to bit ($i + 1$) of the original binary number (this addition is also modulo-2). Thus defining n as a string of binary bits,

$$n = (b_p, b_{p-1}, \ldots, b_1)_2 \qquad (1.40)$$

and expressing this in the Gray code, we write

$$n = (g_p, g_{p-1}, \ldots, g_1)_2 \qquad (1.41)$$

where

$$g_i = b_i \oplus b_{i+1} \qquad (1.42)$$

and $\oplus$ represents modulo-2 addition.

Hence, to find WAL(9, t) we first express 9 in binary code as 1001 and then rearrange this in Gray code as 1101. The second bit position gives $b_i = 0$, so that we can write directly from Eq. (1.37)

$$\text{WAL}(9, t) = \text{RAD}(4, t)\,\text{RAD}(3, t)\,\text{RAD}(1, t) \qquad (1.43)$$

which is the same result as found earlier for PAL(13, t). A graphic illustration of this function product derivation is shown in Fig. 1.13. This shows clearly the role of the function RAD(1, t) in the sign inversion occurring at the centre of the generated function WAL(9, t).

(iii) *Through the Hadamard matrix* The Hadamard matrix is a square array whose coefficients comprise only $+1$ and -1 and in which the rows (and columns) are orthogonal to one another. In a symmetrical Hadamard matrix it is possible to interchange rows and columns or to change the sign of every element in a row without affecting these orthogonal properties. This

TABLE 1.2. Gray single-digit change code.

Decimal	Code	Decimal	Code
0	0000	8	1100
1	0001	9	1101
2	0011	10	1111
3	0010	11	1110
4	0110	12	1010
5	0111	13	1011
6	0101	14	1001
7	0100	15	1000

1.3 The Walsh Function Series

FIG. 1.13. Derivation of WAL(9, *t*) from three Rademacher functions.

makes it possible to obtain a symmetrical Hadamard matrix whose first row and first column contain only +1's. The matrix obtained in this way is known as the *normal form* for the Hadamard matrix. The lowest-order Hadamard matrix is of the order two, viz.,

$$\mathbf{H}_2 = \begin{bmatrix} 1 & 1 \\ 1 & -1 \end{bmatrix} \tag{1.44}$$

Higher-order matrices, restricted to having powers of two, can be obtained from the recursive relationship

$$\mathbf{H}_N = \mathbf{H}_{N/2} \otimes \mathbf{H}_2 \tag{1.45}$$

where $\otimes$ denotes the direct or *Kronecker product* and N is a power of two.

The Kronecker product means replacing each element in the matrix (in this case $\mathbf{H}_{N/2}$) by the matrix $\mathbf{H}_2$. Thus, for

$$\mathbf{H}_4 = \mathbf{H}_2 \otimes \mathbf{H}_2$$

we replace each of the 1's and the -1's of the matrix given above in Eq. (1.44) by the complete matrix of $\mathbf{H}_2$ or its inverse, thus

$$\mathbf{H}_4 = \begin{bmatrix} 1 & 1 & 1 & 1 \\ 1 & -1 & 1 & -1 \\ 1 & 1 & -1 & -1 \\ 1 & -1 & -1 & 1 \end{bmatrix} \tag{1.46}$$

Furthermore, if we now replace each element in the $\mathbf{H}_4$ matrix by an $\mathbf{H}_2$ matrix we obtain an $\mathbf{H}_8$ matrix. By replacing each row of this matrix by its equivalent naturally ordered Walsh function we can form a series of functions which will indicate the ordering obtained through this derivation. Therefore, for a series consisting of eight terms

$$\mathbf{H}_8 = \mathbf{H}_4 \times \mathbf{H}_2 = \begin{bmatrix} 1 & 1 & 1 & 1 & 1 & 1 & 1 & 1 \\ 1 & -1 & 1 & -1 & 1 & -1 & 1 & -1 \\ 1 & 1 & -1 & -1 & 1 & 1 & -1 & -1 \\ 1 & -1 & -1 & 1 & 1 & -1 & -1 & 1 \\ 1 & 1 & 1 & 1 & -1 & -1 & -1 & -1 \\ 1 & -1 & 1 & -1 & -1 & 1 & -1 & 1 \\ 1 & 1 & -1 & -1 & -1 & -1 & 1 & 1 \\ 1 & -1 & -1 & 1 & -1 & 1 & 1 & -1 \end{bmatrix} \equiv \begin{bmatrix} \mathrm{HAD}(0, t) \\ \mathrm{HAD}(1, t) \\ \mathrm{HAD}(2, t) \\ \mathrm{HAD}(3, t) \\ \mathrm{HAD}(4, t) \\ \mathrm{HAD}(5, t) \\ \mathrm{HAD}(6, t) \\ \mathrm{HAD}(7, t) \end{bmatrix} \tag{1.47}$$

The relationship between these Hadamard matrices and a sampled set of Walsh functions is now clear. They simply express a Walsh function series having positive phasing and arranged in natural or Kronecker order.

If the rows of a Hadamard matrix $\mathbf{H}_N$ are rearranged in ascending order of sequency, this becomes a Walsh matrix $\mathbf{W}_N$ having positive phasing as shown below in Eq. (1.48):

$$\mathbf{W}_N = \begin{bmatrix} 1 & 1 & 1 & 1 & 1 & 1 & 1 & 1 \\ 1 & 1 & 1 & 1 & -1 & -1 & -1 & -1 \\ 1 & 1 & -1 & -1 & -1 & -1 & 1 & 1 \\ 1 & 1 & -1 & -1 & 1 & 1 & -1 & -1 \\ 1 & -1 & -1 & 1 & 1 & -1 & -1 & 1 \\ 1 & -1 & -1 & 1 & -1 & 1 & 1 & -1 \\ 1 & -1 & 1 & -1 & -1 & 1 & -1 & 1 \\ 1 & -1 & 1 & -1 & 1 & -1 & 1 & -1 \end{bmatrix} \equiv \begin{bmatrix} \text{WAL}(0, t) \\ \text{SAL}(1, t) \\ \text{CAL}(1, t) \\ \text{SAL}(2, t) \\ \text{CAL}(2, t) \\ \text{SAL}(3, t) \\ \text{CAL}(3, t) \\ \text{SAL}(4, t) \end{bmatrix} \quad (1.48)$$

(iv) *By the use of Boolean synthesis* For Boolean synthesis we use the expression developed by Gibbs [19] which defines the discrete function WAL(n, t) as a form of the continued product definition

$$\text{WAL}(n, t) = -1^{\sum_{i=1}^{p}(n_i + n_{i+1})t_i} \quad (1.49)$$

where $n = 0, 1, \ldots, N - 1$ and $0 \geq t \geq 1$. Defining the number of terms N as 2^p, we can define n in terms of binary bits as

$$n = (n_p, n_{p-1}, \ldots, n_1)_2$$

Similarly we can define t as a binary expansion, viz.,

$$t = (t_1, t_2, \ldots, t_i)_2$$

Since the function WAL(n, t) is bounded by p terms beyond which $n_i = 0$, only the first p binary bits of the t expansion appear in Eq. (1.49), so that by substituting as before a binary 0 for 1 and a binary 1 for -1 we can express WAL(n, t) as an N-bit string, viz.,

$$\text{WAL}(n, t) = \sum_{i=1}^{p} (n_i \oplus n_{i+1}) T_{p-i+1} \quad (1.50)$$

where T is an integer constructed by taking the first p binary digits of t.

Since the Walsh function is symmetric [Eq. (1.23)] we can also write

$$\text{WAL}(T, n) = \sum_{i=1}^{p} (T_i \oplus T_{i+1}) n_{p-i+1} \quad (1.51)$$

A complete set of N Walsh functions can thus be represented by an N by N Boolean matrix using Eq. (1.50) or (1.51).

1.3 The Walsh Function Series

The expression within the parentheses in Eq. (1.51) represents a binary-to-Gray-code conversion defined in Eq. (1.42). Thus Eqs. (1.50) and (1.51) may be rewritten as

$$\text{WAL}(n, T) = \sum_{i=1}^{p} g_i^n T_{p-i+1} \quad (1.52)$$

and

$$\text{WAL}(T, n) = \sum_{i=1}^{p} g_i^T n_{p-i+1} \quad (1.53)$$

where g_i^T is the ith bit of the Gray code,

$$g(T) = (g_p^n, g_{p-1}^n, \ldots, g_1^n)_2 \quad (1.54)$$

This form of derivation is useful in considering hardware function generation (Chapter 4) since the process of modulo-2 addition corresponds to a Boolean exclusive-OR operation whilst multiplication corresponds to a Boolean AND operation. Thus direct implementation of the preceding mathematical expressions can be made by using logical circuits.

1.3.5 Relationships between ordered series

The WAL-, PAL- and HAD-ordered Walsh functions are related (a) through a bit reversal for the position of each component in a series, (b) through a conversion using a Gray code or (c) by a combination of both of these. For example, given a function numbered in dyadic ordering, the corresponding sequency order is given by

$$\text{PAL}(n, t) = \text{WAL}[b(n), t] \quad (1.55)$$

where $b(n)$ is a Gray-code-to-binary conversion of n. A list of converted values for $N = 8$ is given in Table 1.3. A procedure for carrying out this conversion is described by Yuen [22].

The converse relationship,

$$\text{WAL}(n, t) = \text{PAL}(g(n), t) \quad (1.56)$$

As shown earlier in Eqs. (1.40–1.42) we can express the decimal order number n_i in terms of a modulo-2 equivalent b_i. Thus for $n = 6_{10} = 0110_2 = b_3, b_2, b_1, b_0$ then digit

$$g_0(n) = 0 \oplus 1 = 1$$
$$g_1(n) = 1 \oplus 1 = 0$$
$$g_2(n) = 1 \oplus 0 = 1$$

TABLE 1.3. Conversion of a PAL-ordered series to WAL order for $N = 8$.

Decimal	Binary	Gray code	Converted Gray code	Decimal
PAL 0	000	000	000	WAL 0
PAL 1	001	001	001	WAL 1
PAL 2	010	011	011	WAL 3
PAL 3	011	010	010	WAL 2
PAL 4	100	110	111	WAL 7
PAL 5	101	111	110	WAL 6
PAL 6	110	101	100	WAL 4
PAL 7	111	100	101	WAL 5

so that $g(n) = 101_2 = 5_{10}$ and we can write

$$\text{WAL}(6, t) = \text{PAL}(5, t)$$

To find the corresponding sequency order for a function given in natural order we use

$$\text{HAD}(n, t) = \text{WAL}(b(u), t) \tag{1.57}$$

where u represents a bit reversal for n and $b(u)$ is a Gray-code-to-binary conversion of u. Table 1.4 gives bit-reversed and converted values for $N = 8$.

Again the converse relationship

$$\text{WAL}(n, t) = \text{HAD}(g(u), t) \tag{1.58}$$

is easier to calculate and is simply the modulo-2 addition of the bit-reversed values for n as given in Eq. (1.55).

As stated earlier the relationship between dyadic and natural ordering is dependent on bit-reversal of the binary value of the order n. Thus dyadic

TABLE 1.4. Conversion of a HAD-ordered series to WAL order for $N = 8$.

Decimal	Binary	Bit reversal	Converted Gray code	Decimal
HAD 0	000	000	000	WAL 0
HAD 1	001	100	111	WAL 7
HAD 2	010	010	011	WAL 3
HAD 3	011	110	100	WAL 4
HAD 4	100	001	001	WAL 1
HAD 5	101	101	110	WAL 6
HAD 6	110	011	010	WAL 2
HAD 7	111	111	101	WAL 5

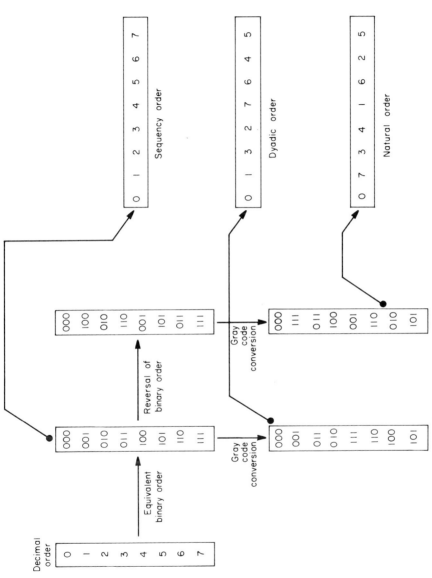

FIG. 1.14. Relationships between three methods of ordering the Walsh function series.

order in terms of natural order is given as

$$\text{PAL}(n, t) = \text{HAD}(u, t) \qquad (1.59)$$

where u represents a bit reversal for n.

Thus for $N = 16$ and $n = 10_{10} = 1010_2$ reversing the bit positions for n gives $0101 = 5_{10}$ and we can write

$$\text{PAL}(10, t) = \text{HAD}(5, t).$$

These interrelationships are summarised in Fig. 1.14 in which both dyadic and natural ordering are related to a linearly progressing sequency order.

Tables 1.5–1.7 provide complete progressive conversion values in terms of each of the three orders for $N = 16$. Note that due to the bit reversal contained in Eqs. (1.57) and (1.58) the order for the corresponding sequency functions to natural functions given in the tables will be different for different values of N.

1.3.6 Relationship between Fourier and Walsh series

From Eqs. (1.24–1.27) it can be seen that for a normalised series, since $\text{WAL}(0, t) = 1$ inside the interval $0 \leq t \leq 1$, the expressions for $a_0/2$ shown for the Fourier and Walsh series are equivalent and give the mean value of the function. Also, if we take Eq. (1.29) to its limit by extending the summa-

TABLE 1.5. Conversion for sequency linear progression ($N = 16$).

Sequency	Dyadic	Natural
WAL(0, t)	PAL(0, t)	HAD(0, t)
WAL(1, t)	PAL(1, t)	HAD(8, t)
WAL(2, t)	PAL(3, t)	HAD(12, t)
WAL(3, t)	PAL(2, t)	HAD(4, t)
WAL(4, t)	PAL(6, t)	HAD(6, t)
WAL(5, t)	PAL(7, t)	HAD(14, t)
WAL(6, t)	PAL(5, t)	HAD(10, t)
WAL(7, t)	PAL(4, t)	HAD(2, t)
WAL(8, t)	PAL(12, t)	HAD(3, t)
WAL(9, t)	PAL(13, t)	HAD(11, t)
WAL(10, t)	PAL(15, t)	HAD(15, t)
WAL(11, t)	PAL(14, t)	HAD(7, t)
WAL(12, t)	PAL(10, t)	HAD(5, t)
WAL(13, t)	PAL(11, t)	HAD(13, t)
WAL(14, t)	PAL(9, t)	HAD(9, t)
WAL(15, t)	PAL(8, t)	HAD(1, t)

1.3 The Walsh Function Series

TABLE 1.6. Conversion for dyadic linear progression ($N = 16$).

Dyadic	Sequency	Natural
PAL(0, t)	WAL(0, t)	HAD(0, t)
PAL(1, t)	WAL(1, t)	HAD(8, t)
PAL(2, t)	WAL(3, t)	HAD(4, t)
PAL(3, t)	WAL(2, t)	HAD(12, t)
PAL(4, t)	WAL(7, t)	HAD(2, t)
PAL(5, t)	WAL(6, t)	HAD(10, t)
PAL(6, t)	WAL(4, t)	HAD(6, t)
PAL(7, t)	WAL(5, t)	HAD(14, t)
PAL(8, t)	WAL(15, t)	HAD(1, t)
PAL(9, t)	WAL(14, t)	HAD(9, t)
PAL(10, t)	WAL(12, t)	HAD(5, t)
PAL(11, t)	WAL(13, t)	HAD(13, t)
PAL(12, t)	WAL(8, t)	HAD(3, t)
PAL(13, t)	WAL(9, t)	HAD(11, t)
PAL(14, t)	WAL(11, t)	HAD(7, t)
PAL(15, t)	WAL(10, t)	HAD(15, t)

tion to infinity, the Fourier and Walsh series representations are identical. Hence, we can substitute the extended version of Eq. (1.29) in the expression for the cosine coefficient given in Eq. (1.25) to obtain

$$a_k = \frac{2}{T} \int_0^T \left[\frac{a_0}{2} + \sum_{i=1}^{\infty} \sum_{j=1}^{\infty} a_i \, \text{SAL}(i, t) + b_j \, \text{CAL}(j, t) \right] \cos k\omega_0 t \, dt \quad (1.60)$$

TABLE 1.7. Conversion for natural linear progression ($N = 16$).

Natural	Sequency	Dyadic
HAD(0, t)	WAL(0, t)	PAL(0, t)
HAD(1, t)	WAL(15, t)	PAL(8, t)
HAD(2, t)	WAL(7, t)	PAL(4, t)
HAD(3, t)	WAL(8, t)	PAL(12, t)
HAD(4, t)	WAL(3, t)	PAL(2, t)
HAD(5, t)	WAL(12, t)	PAL(10, t)
HAD(6, t)	WAL(4, t)	PAL(6, t)
HAD(7, t)	WAL(11, t)	PAL(14, t)
HAD(8, t)	WAL(1, t)	PAL(1, t)
HAD(9, t)	WAL(14, t)	PAL(9, t)
HAD(10, t)	WAL(6, t)	PAL(5, t)
HAD(11, t)	WAL(9, t)	PAL(13, t)
HAD(12, t)	WAL(2, t)	PAL(3, t)
HAD(13, t)	WAL(13, t)	PAL(11, t)
HAD(14, t)	WAL(5, t)	PAL(7, t)
HAD(15, t)	WAL(10, t)	PAL(15, t)

Reversing the order of integration,

$$a_k = a_0 + \frac{2}{T}\sum_{i=1}^{\infty} a_i \left[\int_0^T \text{SAL}(i, t) \cos k\omega_0 t \, dt\right]$$
$$+ \frac{2}{T}\sum_{j=1}^{\infty} b_j \left[\int_0^T \text{CAL}(j, t) \cos k\omega_0 t \, dt\right] \quad (1.61)$$

The terms in brackets represent the Fourier coefficients for SAL(i, t) and CAL(j, t), respectively. This can be seen if we substitute SAL(i, t) and CAL(j, t) for $x(t)$ in the cosine Eq. (1.25). Writing a_k(SAL) and a_k(CAL) for these we have

$$a_k = a_0 + \frac{2}{T}\sum_{j=1}^{\infty}\sum_{i=1}^{\infty}\left[a_i a_k(\text{SAL}) + b_j a_k(\text{CAL})\right] \quad (1.62)$$

Similarly we can derive for b_k

$$b_k = a_0 + \frac{2}{T}\sum_{j=1}^{\infty}\sum_{i=1}^{\infty}\left[a_i b_k(\text{SAL}) + b_j b_k(\text{CAL})\right] \quad (1.63)$$

To find the Fourier series in terms of the Walsh series the terms for a_k and b_k are substituted in Eq. (1.24) to give

$$x(t) = \frac{a_0}{2} + \sum_{k=1}^{\infty}\left\{\left[a_0 + \frac{2}{T}\sum_{j=1}^{\infty}\sum_{i=1}^{\infty}[a_i a_k(\text{SAL}) + b_j a_k(\text{CAL})]\cos k\omega_0 t\right]\right.$$
$$\left.+ \left[a_0 + \frac{2}{T}\sum_{j=1}^{\infty}\sum_{i=1}^{\infty}[a_i b_k(\text{SAL}) + b_j b_k(\text{CAL})]\sin k\omega_0 t\right]\right\} \quad (1.64)$$

Since both Walsh and Fourier series are orthogonal, the terms containing a_0, $a_i a_k$(SAL) and $b_j b_k$(CAL) will vanish so that Eq. (1.64) simplifies to

$$x(t) = \frac{a_0}{2} + \frac{2}{T}\sum_{k=1}^{\infty}\sum_{i=1}^{\infty}\sum_{j=1}^{\infty}$$
$$\times [a_i b_k(\text{SAL})\sin(k\omega_0 t) + b_j a_k(\text{CAL})\cos(k\omega_0 t)] \quad (1.65)$$

which approximates to a limited and normalised sampled form

$$x(t) = \frac{a_0}{2} + \frac{2}{N}\sum_{k=1}^{N}\sum_{i=1}^{N}\sum_{j=1}^{N}$$
$$\times \left[a_i b_k(\text{SAL})\sin\left(2\pi k\frac{i}{N}\right) + b_j a_k(\text{CAL})\cos\left(2\pi k\frac{i}{k}\right)\right],$$
$$k, i, j = 0, 1, \ldots, N-1 \quad (1.66)$$

1.3 The Walsh Function Series

Equation (1.66) enables the Fourier series for $x(t)$ to be obtained from the Walsh series by using a table of Fourier coefficients for the Walsh functions SAL(i, t) and CAL(j, t). This method of derivation for the Fourier coefficients is used in hardware generation by Siemens and Kitai [23], who give examples of coefficient tables for common types of waveforms.

1.3.7 Relationship between CAL and SAL functions

Although a simple relationship exists between $\cos(2\pi kt)$ and $\sin(2\pi kt)$, the relationship between CAL(k, t) and SAL(k, t) is rather complicated. This is a consequence of the absence of a simple shift theorem which is found in the circular functions.

A relationship has been stated by Pichler [24] and modified by Tam and Goulet [25] to give the expression

$$\text{CAL}(k, t + t_0) = \text{SAL}(k, t) \qquad (1.67)$$

where

$$t_0 = (-1)^{q+1} 2^{-(r+2)} \qquad (1.68)$$

For a given value of k the time lag t_0 between CAL and SAL functions of the same order can be derived through a set of pairs of values q and r. These are expressed uniquely by a single factor k, viz.,

$$k = 2^r(2q + 1), \qquad r, q = 0, 1, 2, \ldots \qquad (1.69)$$

A table for increasing values of k in terms of r and q can be obtained from a recursive implementation of Eq. (1.69). These values are shown in Table 1.8 for the first eight values of k.

TABLE 1.8. Relationship between CAL and SAL functions.

k	r	q
1	0	0
2	1	0
3	1	0
4	2	0
5	0	2
6	1	1
7	0	3
8	3	0

1.4 The Haar function series

The set of Haar functions form an orthogonal system of periodic square waves taking the values of 1, 0 and -1 multiplied by powers of $\sqrt{2}$.

A recurrence relation which can be used to generate a set of such functions within the interval $0 \leq t \leq 1$ is given by

$$\text{HAR}(0, 0, t) = 1$$

$$\text{HAR}(p, r, t) = \begin{cases} 2^{p/2} & \text{for } (r - 1/2^p) \leq t \leq (r - \frac{1}{2}/2^p) \\ -2^{p/2} & \text{for } (r - \frac{1}{2}/2^p) \leq t \leq (r/2^p) \\ 0 & \text{elsewhere} \end{cases} \quad (1.70)$$

where $p = 0, 1, 2, \ldots r = 1, 2, \ldots, 2^p (= N)$. This enables us to refer to the Haar functions by *order r* and *degree p* as well as time t. The degree p then denotes a subset having the same number of zero crossings in a given width $1/2^p$, thus providing a form of comparison with frequency and sequency terminology. The order r gives the position of the function within this subset. All members of the subset with the same degree are obtained by shifting the first member along the axis by an amount proportional to its order.

A simplified definition proposed by Kremer [26] replaces r by $m + 1$, and we can write

$$\text{HAR}(0, t) = 1 \qquad \text{for } 0 \leq t \leq 1$$

$$\text{HAR}(1, t) = \begin{cases} 1 & \text{for } 0 \leq t < \frac{1}{2} \\ -1 & \text{for } \frac{1}{2} \leq t \leq 1 \end{cases}$$

$$\text{HAR}(2, t) = \begin{cases} \sqrt{2} & \text{for } 0 \leq t < \frac{1}{4} \\ -\sqrt{2} & \text{for } \frac{1}{4} \leq t < \frac{1}{2} \\ 0 & \text{for } \frac{1}{2} \leq t \leq 1 \end{cases}$$

$$\text{HAR}(3, t) = \begin{cases} 0 & \text{for } 0 \leq t < \frac{1}{2} \\ \sqrt{2} & \text{for } \frac{1}{2} \leq t < \frac{3}{4} \\ -\sqrt{2} & \text{for } \frac{3}{4} \leq t \leq 1 \end{cases} \quad (1.71)$$

$$\vdots$$

(continued)

1.4 The Haar Function Series

$$\text{HAR}(2^p + m, t) = \begin{cases} (\sqrt{2})^p & \text{for } m/2^p \leq t < (m + \tfrac{1}{2})2^p \\ -(\sqrt{2})^p & \text{for } (m + \tfrac{1}{2})/2^p \leq t < (m + 1)/2^p \\ 0 & \text{for elsewhere} \end{cases}$$

where $p = 0, 1, 2, \ldots, m = 0, 1, \ldots, 2^p - 1$. This allows a sequential numbering system analogous to that adopted by Walsh for this function series and is known as *rank order*.

The first eight Haar functions ordered in this way were shown in Fig. 1.5. The first two functions are identical to WAL(0, t) and WAL(1, t). The next function, HAR(2, t) is simply HAR(1, t) squeezed into the left-hand half of the time base and modified in amplitude to $\pm\sqrt{2}$. The next function, HAR(3, t), is identical but squeezed into the right-hand half of the time base. Subsequent pairs of functions are similarly squeezed and shifted, having amplitudes of ± 1 multiplied by powers of $\sqrt{2}$. In general all members of the same function subset (such as HAR(2, t) and HAR(3, t) or HAR(4, t), HAR(5, t), HAR(6, t) and HAR(7, t), etc.) are obtained by a lateral shift of the first member along the time axis by an amount proportional to its length.

An alternative order known as *natural order* is also found. Here the series is derived from Kronecker products in a similar way to that described earlier for Walsh functions. This is computationally simpler but requires a *zonal bit reversal* procedure after the generation to obtain rank ordering [27].

From Fig. 1.5, the essential characteristic of the Haar function is seen as a constant value everywhere except in one sub-interval where a double step occurs. This type of function is also found in a single line scan for certain types of images and has led to the suggestion that the Haar function can be useful in edge detection for pattern recognition [28].

The orthogonal nature of the Haar function can be seen from Eq. (1.70), and the proof of completeness of the series is given by Haar [5].

A given continuous function $x(t)$ within the interval $0 \leq t \leq 1$ and repeated periodically outside this interval can be synthesised from a Haar series by

$$x(t) = \sum_{n=0}^{\infty} C_n \, \text{HAR}(n, t) \qquad (1.72)$$

where

$$C_n = \int_{t=0}^{1} x(t) \, \text{HAR}(n, t) \, dt \qquad (1.73)$$

As with other complete series, *Parseval's equivalence* holds, and we can write

$$\int_{t=0}^{1} x^2(t) \, dt = \sum_{n=0}^{\infty} C_n^2 \qquad (1.74)$$

There is no theorem analogous to the shift or addition theorems found with Fourier and Walsh series, respectively.

The convergence features of the expansion in Haar function are superior to the Walsh functions, as noted by Alexits [29], since for some continuous functions the Walsh expansion can actually diverge at a given point. This cannot occur in the case of Haar expansions. Any finite approximation to a function $x(t)$ using the Haar series will take the form of a step function having 2^p equal-length steps.

If we consider the value of a partial sum at each step size, we find that this is simply the mean value of $x(t)$ in the interval covered by the step. This is the condition for the best step function approximation of $x(t)$ in terms of the MSE and accounts for the comparatively small number of terms necessary to synthesise a waveform from the Haar series [30].

1.5 Mixed function series

Provided that a function series is orthogonal and complete, it may be used for accurate synthesis of a given signal from a limited number of function elements and its transform will generally provide some measure of data compression. How efficiently the series carries out these roles can depend to a large extent on the nature of the signal and how well the function elements correspond with its individual characteristics. This is particularly important for digital image processing in which a very large number of discrete sampled values are involved and it is desirable to limit the number of significant constituent or transformed values.

The realisation of these factors together with the availability of low-cost digital hardware has stimulated the design and application of a number of sequency function series and transformation algorithms for specific purposes. Some of these represent combinations of the well-known sequency functions considered earlier in this chapter and are known as hybrid series. Others introduce new functions such as the slant functions into an existing sequency series.

1.5.1 Hybrid series

A hybrid version of the Haar and Walsh transform has been developed by Rao *et al.* [31]. This combines the advantages of both functions and is valuable for feature selection and pattern recognition [32]. One hybrid series resulting from this is shown in Fig. 1.15 for $N = 8$.

The series may be derived from linear combinations of Haar functions

1.5 Mixed Function Series

and an appropriate scaling factor. Thus denoting the functions as HAW(n, t) we can write for Fig. 1.15

$$HAW(0, t) = HAR(0, t)$$

$$HAW(1, t) = \frac{1}{\sqrt{2}}(HAR(2, t) + HAR(3, t))$$

$$HAW(2, t) = \frac{\sqrt{2}}{2}(HAR(4, t) + HAR(6, t))$$

$$HAW(3, t) = \frac{\sqrt{2}}{2}(HAR(5, t) + HAR(7, t))$$

$$HAW(4, t) = HAR(1, t) \quad (1.75)$$

$$HAW(5, t) = \frac{1}{\sqrt{2}}(HAR(2, t) - HAR(3, t))$$

$$HAW(6, t) = \frac{\sqrt{2}}{2}(HAR(4, t) - HAR(6, t))$$

$$HAW(7, t) = \frac{\sqrt{2}}{2}(HAR(5, t) - HAR(7, t))$$

A mixed Walsh–Haar series has also been suggested by Huang [33] for use in a fast transformation matrix enabling elements of both Walsh and Haar functions to be derived in a single operation. It is derived from a matrix expansion of a Hadamard and a scaled unitary matrix and consists of the first $N/2$ sequency Walsh functions followed by $N/2$ Haar functions, excluding those which are common with the Walsh series (i.e., WAL(0, t) and WAL(1, t)).

1.5.2 Slant series

This series of functions, developed specifically for image transmission purposes [6], is related to the Walsh function series WAL(n, t), where this is expressed in terms of positive phasing (see Subsection 1.3.3).

The relationship is best considered through a transformation matrix. For a series of length $N = 2$ the slant matrix S_2 is identical to the Hadamard matrix which we considered in (1.3.4), viz.,

$$\mathbf{S}_2 = \frac{1}{\sqrt{2}}\begin{bmatrix} 1 & 1 \\ 1 & -1 \end{bmatrix} = k\mathbf{H}_2 \quad (1.76)$$

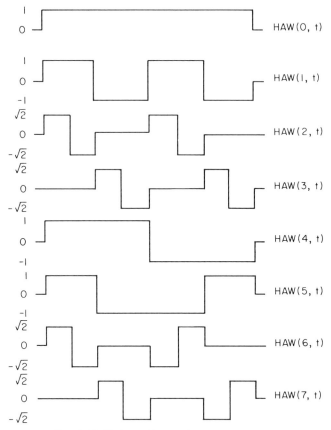

FIG. 1.15. Hadamard–Haar functions for $N = 8$.

The slant matrix for $N = 4$ can be written as

$$\mathbf{S}_4 = \frac{1}{\sqrt{4}} \begin{bmatrix} 1 & 1 & 1 & 1 \\ a+b & a-b & -a+b & -a-b \\ 1 & -1 & -1 & 1 \\ a-b & -a-b & a+b & -a+b \end{bmatrix} \quad (1.77)$$

The slant formation of the second and fourth rows is achieved by the appropriate signs of the constants a and b.

It can easily be shown that to obtain a uniform step size between series elements and hence linear slope $a = 2b$ and to achieve orthogonality for $\mathbf{S}_4$, b must equal $1/\sqrt{5}$ [34].

1.5 Mixed Function Series

Hence we can write

$$\mathbf{S}_4 = \frac{1}{\sqrt{4}} \begin{bmatrix} 1 & 1 & 1 & 1 \\ \frac{3}{\sqrt{5}} & \frac{1}{\sqrt{5}} & \frac{-1}{\sqrt{5}} & \frac{-3}{\sqrt{5}} \\ 1 & -1 & -1 & 1 \\ \frac{1}{\sqrt{5}} & \frac{-3}{\sqrt{5}} & \frac{3}{\sqrt{5}} & \frac{-1}{\sqrt{5}} \end{bmatrix} \quad (1.78)$$

Here $\mathbf{S}_4$ will be orthonormal and possesses a sequency property; i.e., each row in the series has an increasing number of sign reversals from 0 to 3.

Pratt et al. [34] have developed an extension to this slant matrix for any order $N \times N$, where N is a power of 2, by a form of Kronecker product expansion similar to that described in Subsection 1.3.4. This gives each higher-order matrix of order N in terms of a slant matrix of order $N/2$ and a 2×2 identity matrix.

As would be expected from this derivation, it is possible to obtain an algebraic relationship between the slant and Walsh series through the combination of a number of Walsh functions with suitable weighting to form a new slant function [35]. Thus for $N = 8$ we have

$\text{SLA}(0, t) = \text{WAL}(0, t)$

$\text{SLA}(1, t) = \frac{1}{\sqrt{21}}(4\ \text{WAL}(1, t) + 2\ \text{WAL}(3, t) + \text{WAL}(7, t))$

$\text{SLA}(2, t) = \frac{1}{\sqrt{5}}(2\ \text{WAL}(2, t) + \text{WAL}(6, t))$

$\text{SLA}(3, t) = \frac{1}{\sqrt{21}} \frac{1}{\sqrt{5}}(-5\ \text{WAL}(1, t) + 8\ \text{WAL}(3, t) + 4\ \text{WAL}(7, t))$ \quad (1.79)

$\text{SLA}(4, t) = \text{WAL}(4, t)$

$\text{SLA}(5, t) = \text{WAL}(5, t)$

$\text{SLA}(6, t) = \frac{1}{\sqrt{5}}(-\text{WAL}(2, t) + 2\ \text{WAL}(6, t))$

$\text{SLA}(7, t) = \frac{1}{\sqrt{5}}(-\text{WAL}(3, t) + 2\ \text{WAL}(7, t))$

This can be verified by combining the appropriate functions from Figs. 1.4 and 1.6 and taking note of the scaling constants and the need to invert some of the Walsh functions for positive phasing.

A slant Haar series has also been proposed by Fino and Algazi [36].

1.6 Discrete sampled functions

In the foregoing treatment of sequency functions the signal to be synthesised and the functions themselves were considered as continuous functions of time. In a practical realisation we will be considering sampled and quantised values for digital processing. The *sampling theorem* for sine–cosine functions states that for a band-limited and sampled signal containing frequencies up to f_n Hz, a sampling rate of $2f_n = f_s$ (or sampling period of $h = 1/2f_n = 1/f_s$) is the minimum rate necessary to recover completely the original signal from the sampled version. A similar theorem is applicable to the Walsh series, but the minimum sampling rate may not be $2Z$ for a sequency-limited signal containing sequencies no higher than Zps.

The relevance of sampling theory has been considered by Kak [37], who has shown that the minimum sampling rate should be $f_s = 2^{k+1}$ where the sequency bandwidth is expressed as a power of 2, i.e., $Z = 2^k$ (or sampling period of $h = 1/2^{k+1}$).

Thus, if $Z = 3$ as found in SAL(3, t), then we must put $Z = 4.0$ (the next nearest power of 2), giving $f_s = 2^{2+1} = 8$ rather than $f_s = 2Z = 6$, which would be expected from the sampling theorem for circular functions. Note that for $Z = 4.0$ as found in SAL(4, t) it is permissible to use the same sampling rate of $f_s = 2^{2+1} = 8$.

Aliasing, which is a consequence of the sampling theorem, is equally applicable to the Walsh and Haar functions even though the functions are both time and sequency limited. It is necessary, however, to consider the power-of-2 representation of the sequency bandwidth, as defined above, when determining the cut-off sequency for the band-limiting low-pass sequency filter required.

1.6.1 Effects of quantisation level and sampling interval

Provided that the sampling interval chosen is several times smaller than the Nyquist interval, reconstruction using the Fourier transform results in a linear interpolated version of the original waveform. This is to be expected from the linear properties of the FFT and has been noted elsewhere [38].

In the use of the Walsh transform a limitation is found where the order of the highest Walsh function determines the quantisation level of the reconstructed signal. Let us take, for example, a known case for a threshold level of 2% where only 24 non-zero Walsh coefficients are determined and the highest coefficient found is WAL(128, t). Here $N = 1024$ and the minimum number of reconstructed data points found to represent any quantised level will be $1024/128 = 8$, hence limiting to a known finite value the realisable accuracy of representation.

No such restriction is found with the Fourier transform since a limited time base is not present and a closer approximation to the original function can be obtained.

Where the signal can be related to a physical system having a spring-mass-damped form, synthesis of the waveform from a given number of sinusoidal functions is realisable. This is the case for many physical experiments, and here the use of Fourier, rather than Walsh, transformation is likely to give reduced errors in any form of subsequent analysis of the signal. A rectangle-based signal generally results from technology-generated systems (e.g., a communication coding system) and analysis advantages are obtained if the Walsh or Haar transform is used, particularly if the signal can be binary related to the time base of the series [39]. A possible criteria for the efficiency of reconstruction permitting the selection of the Walsh, Haar or Fourier transform is to be found in a mean-square error technique. Thus, if the original sampled signal x_i is processed with a given threshold level and the reconstructed version from a limited number of coefficients is designated as y_i, then the MSE coefficient can be stated as

$$\text{MSE} = \frac{1}{N}\sum_{i=1}^{N}(x_i - y_i)^2 \tag{1.80}$$

This is not necessarily a valid criteria for two-dimensional data, as will be discussed later.

Risch and Brubaker [40] have investigated the MSE arising from signal reconstruction of a finite discrete signal by using a $(\sin x)/x$ function, a Walsh function and a zero-order hold representation. They show that the $(\sin x)/x$ reconstruction, which is related to a Fourier series, gives the least error when N is small. As the number of sampled points is increased, reconstruction using the Walsh function will give less error for $N > 32$. At all times the error with Walsh reconstruction is less than with zero-order hold.

References

1. Wiener, N. (1949). "Extrapolation, Interpolation and Smoothing of Stationary Time Series." MIT Press, Cambridge, Massachusetts.
2. Rademacher, H. (1922). Einige Sätze von allgemeinen Orthogonalfunktionen. *Math. Annal* **87**, 122–138.
3. Walsh, J. L. (1923). A closed set of orthogonal functions. *Ann. J. Math* **55**, 5–24.
4. Pratt, W. K., Kane, J., and Andrews, H. C. (1969). Hadamard transform image coding. *Proc. IEEE* **57**, 58–68.
5. Haar, A. (1910). Zür Theorie der orthogonalen Funktionensysteme. *Math. Annal.* **69**, 331–71.
6. Enomoto, H., and Shibata, K. (1971). Orthogonal transform coding system for television signals. *Proc. Symp. Applic. Walsh Functions, Washington, D.C., 1971*, AD727000, pp. 11–17.

7. Ahmed, N., and Rao, K. R. (1975). "Orthogonal Transforms for Digital Signal Processing." Springer-Verlag, Berlin.
8. Davis, H. F. (1963). "Fourier Series and Orthogonal Functions." Allyn and Bacon, Boston.
9. Lebedev, N. N. (1965). "Special Functions and Their Applications." Prentice-Hall, Englewood Cliffs, New Jersey.
10. Harmuth, H. F. (1972). "Transmission of Information by Orthogonal Functions," 2nd ed. Springer-Verlag, Berlin.
11. Paley, R. E. (1932). A remarkable set of orthogonal functions. *Proc. London. Math Soc.* **34**, 241–279.
12. Redinbo, G. R. (1971). A note in the construction of generalized Walsh functions. *SIAM J. Math Anal.* **2**, 166–171.
13. Gibbs, J. E. (1969). Walsh functions as solutions of a logical differential equation. National Physical Laboratory, Report DES 1, Teddington, England.
14. Henderson, K. W. (1970). Comment on 'Computation of the fast Walsh–Fourier transform'. *IEEE Trans. Comput.* **C-19**, 850–851.
15. Kunz, H. O., and Ramm-Arnet, J. (1978). Walsh matrices. *Arch. Elek, Ubertragung,* **32**, 56–58.
16. Rao, K. R., Naramsimhan, M. A., and Devarajan, V. (1978). Cal–Sal Walsh Hadamard transform. *IEEE Trans. Acoust. Sp. Sig. Proc.* **ASSP-26**, 605–607.
17. Hurst, S. L. (1978). "Logical Processing of Digital Signals." Crane Russak, New York; Edward Arnold, London.
18. Rushforth, C. K. (1969). Fast Fourier–Hadamard decoding of orthogonal code. *Information Control* **15**, 33–37.
19. Gibbs, J. E. (1970). Discrete complex Walsh transforms. *Proc. Symp. Applic. Walsh Functions, Washington, D.C.,* AD707431, pp. 106–122.
20. Chien, T. M. (1975). On representation of Walsh functions. *IEEE Trans. Electromag. Compat.* **EMC-17**, 170–177.
21. Lackey, R. B., and Meltzer, D. (1971). A simplified definition of the Walsh functions. *IEEE Trans. Comput.* **C-20**, 211–213.
22. Yuen, C. K. (1971). Walsh functions and the Gray code. *Proc. Applic. Walsh Functions, Washington, D.C.,* AD727000, pp. 68–73.
23. Siemens, K. H., and Kitai, R. (1969). Digital Walsh fourier analysis of periodic waveforms. *IEEE Trans. Instrum. Meas.* **IM18**, 316–321.
24. Pichler, F. (1968). Synthese linearer periodische zeitvariabler Filter mit vorgeschriebenen sequenzverhalten. *AEU* **22**, 150–161.
25. Tam, L. D. C., and Goulet, R. Y. (1972). On arithmetic shift for Walsh functions. *IEEE Trans. Comput.* **C21**, 1451–1452.
26. Kremer, H. (1971). Algorithms for the Haar functions and the fast Haar transform. *Proc. Theory Applic. Walsh Functions, Hatfield Polytechnic, England.*
27. Fino, B. J., and Algazi, V. R. (1977). A unified treatment of discrete fast unitary transforms. *SIAM J. Comput.* **6**, 700–717.
28. Rosenfield, A., and Thurston, M. (1971). Edge and curve detection for visual scene analysis. *IEEE Trans. Comput.* **C20**, 562–569.
29. Alexits, G. (1961). "Convergence Problems of Orthogonal Series." Pergamon, New York and London.
30. Shore, J. E. (1973). On the application of Haar functions. *IEEE Trans. Commun.* **COM21**, 209–216.
31. Rao, K. R., Naramsimhan, M. A., and Revulieru, K. (1974). Hadamard–Haar transform. *IEEE Ann. Symp. South Eastern System Theory, 6th, Baton Rouge, Louisiana.*
32. Rao, K. R., Naramsimhan, M. A., and Revulieri, K. (1975). Image data processing by Hadamard–Haar transform. *IEEE Trans. Comput.* **C24**, 888–896.

33. Huang, D. M. (1980). Walsh–Hadamard–Haar hybrid transforms. *IEEE Proc. Int. Conf. Pattern Recognition, 5th,* pp. 180–182.
34. Pratt, W. K., Welch, L. R., and Chen, W. H. (1972). Slant transforms for image coding. *Proc. Applic. Walsh Functions, Washington, D.C.,* AD744650, pp. 229–234.
35. Shibata, K. (1972). Waveform analysis of image signals by orthogonal transformation. *Proc. Applic. Walsh Functions, Washington, D.C.,* AD744650, pp. 210–215.
36. Fino, B. J., and Algazi, V. R. (1974). Slant Haar transform. *Proc. IEEE (Lett.)* **62,** 653–654.
37. Kak, S. C. (1970). Sampling theorem in Walsh–Fourier analysis. *Electron. Lett.* **6,** 14.
38. Cooley, J. W., Lewis, A. W., and Welch, P. D. (1969). The fast Fourier transform and its applications. *IEEE Trans. Ed.* **E12,** 27.
39. Beauchamp, K. G. (1973). Waveform synthesis using Fourier and Walsh series. *Proc. Theory Applic. Walsh Functions, Hatfield Polytechnic, England.*
40. Risch, P. R., and Brubaker, T. A. (1973). Evaluation of data reconstruction using Walsh functions. *Electron. Lett.* **9,** 489–490.

Chapter 2

Transformation

2.1 Introduction

Orthogonal transformation, as a practical tool for processing and analysing signals and images, became possible only with the widespread availability of digital techniques and the digital computer. For any but very small numbers of data values the computational task was, however, quite formidable until the advent of efficient transformation algorithms which applied matrix factoring and partitioning methods to reduce the considerable computational and memory requirements. Initially these algorithms were applied only to discrete Fourier analysis but have now been extended to a large number of other orthogonal transforms either to achieve an even more efficient domain transformation or to take advantage of the special characteristics of these newer transforms.

Orthogonal decomposition is carried out optimally (in the least-squares sense) by calculating the eigenvalues and eigenvectors of a linear system of data. This is how a mathematician would attempt the problem. It is, however, a very difficult if not a numerically impossible task to carry this out for any useful size set of discrete data. Hence for practical and engineering purposes a *sub-optimal solution* is sought, and all of the transforms discussed later fall into this sub-optimal class.

Fortunately for digital computational purposes many sub-optimal transformation techniques are available, all involving fast algorithms and some approaching fairly closely in performance the optimal Karhunen–Loève transform referred to earlier (Subsection 1.2.7). This transform, we recall, is optimum in the sense that it decomposes a time series by packing the most energy into the smallest number of coefficients and minimises the mean-square error between a reconstructed version and the original signal. The class of transformations we will be concerned with are those which are implementable in $pN \log_p N$ or less operations compared with the N^2 operations required for arbitrary linear transformations. These include the Fourier, Hadamard, Walsh, Haar, slant, and various other transforms whose series were discussed in the preceding chapter.

The underlying principle for the efficient implementation of the *fast* discrete transformations carried out on the digital computer is the reduction of the high degree of redundancy present in the transform matrix representation. This was first recognised by Good in 1958 [1] and resulted in the development of fast Fourier transform algorithms by Cooley and Tukey [2], Gentleman and Sande [3] and others.

Since then a number of fast transform algorithms have been described for the range of discrete sequency orthogonal series considered in the preceding chapter. Several of these are based on the fast Fourier transform (FFT), and indeed a generalised transform has been proposed which enables the Fourier, Walsh and other transformations to be determined through an appropriate choice of multiplying constants. Others have been developed differently, and such is the efficiency of some of the newer algorithms that it is possible to arrive at a fast discrete Fourier transformation through other orthogonal transforms at a faster rate than can be obtained through the use of the FFT directly.

A major purpose of this chapter is to consider the essential characteristics of these algorithms and to compare their advantages in various processing and analysis situations. First, however, it is necessary to consider some of the fundamental relationships of these sequency transforms, commencing with the Walsh transform, and in particular to see how these relate to similar characteristics found in the more familiar Fourier transform.

2.2 The discrete Walsh transform

As noted earlier any integrable function $x(t)$ may be represented by a Walsh series defined over the interval (0, 1) as

$$x(t) = a_0 + a_1 \text{WAL}(1, t) + a_2 \text{WAL}(2, t) + \cdots \quad (2.1)$$

where the coefficients are given by

$$a_n = \frac{1}{T} \int_0^T x(t) \, \text{WAL}(n, t) \, dt \quad (2.2)$$

From this we are able to define a transform pair,

$$x(t) = \sum_{k=0}^{\infty} X(n) \, \text{WAL}(n, t) \tag{2.3}$$

$$X(n) = \frac{1}{T} \int_0^T x(t) \, \text{WAL}(n, t) \, dt \tag{2.4}$$

This definition applies to a continuous function limited in time over the interval $0 \geq t \geq 1$. For numerical use it is convenient to consider a discrete series of N terms set up by sampling the continuous functions at N equally spaced points over the interval $(0, 1)$. In order that the properties of the continuous and discrete systems should correspond, we must make N equal to a power of 2, i.e., $N = 2^p$.

The integration shown in Eq. (2.4) may then be replaced by summation through using the trapezium rule of N sampling points x_i, and we can write the *finite discrete Walsh transform pair* as

$$X_n = \frac{1}{N} \sum_{i=0}^{N-1} x_i \, \text{WAL}(n, i), \qquad n = 0, 1, 2, \ldots, N-1 \tag{2.5}$$

and

$$x_i = \sum_{n=0}^{N-1} X_n \, \text{WAL}(n, i), \qquad i = 0, 1, 2, \ldots, N-1 \tag{2.6}$$

Similar transforms $X_c(k)$ and $X_s(k)$ can be obtained for a time series x_i by using Harmuth's CAL and SAL functions, viz.,

$$X_c(k) = \frac{1}{N} \sum_{i=0}^{N-1} x_i \, \text{CAL}(k, i) \tag{2.7}$$

and

$$X_s(k) = \frac{1}{N} \sum_{i=0}^{N-1} x_i \, \text{SAL}(k, i) \tag{2.8}$$

where $k = 1, 2, \ldots, N/2$.

As with their functions the transforms are linear so that if

$$x_i \leftrightarrow X_n \quad \text{and} \quad y_i \leftrightarrow Y_n \quad \text{then} \quad ax_i + by_i \leftrightarrow aX_n + bY_n$$

where a and b are real constants and $\leftrightarrow$ denotes a *transform operator*.

Also the transform is symmetrical (Eq. (1.23)). Since $\text{WAL}(n, i)$ is symmetric about the mid-point of the sequence $i = 0, 1, 2, \ldots, N-1$ when n is even and anti-symmetric when n is odd, it also follows from Eq. (1.22) that a sequence x_i will have a transform composed only of even-order Walsh function coefficients (CAL function) if it is symmetric about its midpoint and be composed only of odd-order (SAL function) coefficients if the series is inversely symmetric.

2.2.1 Comparison with the discrete Fourier transform

The transform and its inverse as given by Eqs. (2.5) and (2.6) may be obtained by matrix multiplication using the digital computer. Since the matrices are symmetric for the DWT (unlike for the DFT), both transform and inverse transform are identical except for a scaling factor $1/N$.

If we compare Eq. (2.5) with the corresponding discrete Fourier transform

$$X_f = \frac{1}{N} \sum_{i=0}^{N-1} x_i \exp(-j2\pi i f/N)$$

$$f = 0, 1, 2, \ldots, N-1, \quad i = 0, 1, 2, \ldots, N-1 \quad (2.9)$$

we note that whilst $\text{WAL}(n, i)$ is real and limited to values ± 1 the kernel $K = \exp(-j2\pi i f/N)$ is complex and can assume N different values for each coefficient. As a direct consequence of this, the Walsh transform proves considerably easier and faster to calculate by digital methods.

2.2.2 Effects of circular time shift

The Fourier and Walsh transforms behave quite differently in response to a change in the phase of the input signal. Consider a discrete signal x_i where the first value is shifted by a finite value h to become x_{i+h} and subsequent terms increment up to x_{N-1} and are then followed by terms x_0 to x_{h-1} to complete the finite series of N terms. This is known as a *circularly-shifted* signal and is equivalent to a phase shift of the continuous input signal x_i ($i = 0, 1, \ldots, N-1$).

The discrete Fourier transform of this shifted signal is given as

$$X'_f = X_f \exp(2\pi f h/N) \quad (2.10)$$

and only a change in amplitude (scaling) takes place—the transform frequency spectrum remains unaltered; i.e., the transform is an invarient one. This is not the case with the Walsh transform [4], so that

$$X'_w \neq kX_w \quad (2.11)$$

where k is a constant.

However, Pichler [5] has shown that if the time shift is obtained through a dyadic translation

$$z_i = x_{i \oplus p} \quad (2.12)$$

(where $\oplus$ indicates modulo-2 addition for the binary representations of i and p), then

$$Z_c^2(k, i) + Z_s^2(k, i) = X_c^2(k, i) + X_s^2(k, i) \quad (2.13)$$

and the sequency spectrum is invariant under the dyadic time shift of the input signal. This also indicates that a relationship must exist between the sequency values obtained, namely,

$$Z(k, i) = X(k \oplus p, i) \qquad (2.14)$$

and is shown in the following example.

Table 2.1 shows the DWT of a single cycle of a sampled sinusoidal waveform having 32 values, which is compared with a transform of the same waveform circularly-shifted through $\pi/2$ rad. If we take the shift index p as $p = \log_2 8 = 3$, then we obtain the following index values for the sequency coefficients k and $k \oplus p$:

90° shift (k)	Added binary values of k and p			0° shift ($k \oplus p$)
1	0001	+00011	= 0010	2
2	0010	:	= 0001	1
5	0101	:	= 0110	6
6	0110	:	= 0101	5
10	1010	:	= 1001	9
13	1101	:	= 1110	14
14	1110	:	= 1101	13
18	10010	:	= 10001	17
26	11010	:	= 11001	25
30	11110	:	= 11101	29

Taking respective values from Table 2.1 we obtain,

| 90° shift | | 0° shift | |
k	value	$k \oplus p$	value
1	−0.063	2	0.063
2	0.633	1	0.633
5	0.025	6	0.025
6	0.263	5	−0.263
10	−0.052	9	−0.052
13	0.013	14	0.013
14	0.126	13	−0.126
18	−0.013	17	−0.013
26	−0.025	25	−0.025
30	0.062	29	−0.062

Squaring these values will give the expected invariance.

Figure 2.1 shows the effect of circular phase shift on a sampled sinusoidal signal. Two features of interest may be seen from this diagram. The value of the sequency coefficients changes quite considerably over a 2π-rad phase shift and a change of sign also occurs. It may be noted further that complementary changes in value occur for related pairs of CAL and SAL coefficients. As a consequence the effect of circular phase shift is of less importance where the sum of the squares of pairs of transformed coefficients of the same

TABLE 2.1 (a). Walsh transform coefficients for a simple sine waveform, $N = 32$.

0	0.663	0.063	0	0	−0.263	0.025	0
0	−0.052	−0.006	0	0	−0.126	0.013	0
0	−0.013	−0.002	0	0	0.006	0	0
0	−0.025	−0.002	0	0	−0.062	0.006	0

TABLE 2.1 (b). Walsh transform coefficients for the sine waveform shifted by 90°.

0	−0.063	0.663	0	0	0.025	0.263	0
0	0.006	−0.052	0	0	0.013	0.126	0
0	0.001	−0.013	0	0	0	−0.006	0
0	0.002	−0.025	0	0	0.006	0.062	0

frequency are taken, as in power spectrum derivation. Nevertheless, for many applications this effect is inconvenient, and in a later section a number of invariant sequency transforms which do not exhibit this phase shifting will be considered.

2.2.3 Behaviour of transform products

The behaviour of transform products for Walsh functions is determined from an *addition relationship* [6]

$$\text{WAL}(n, t)\,\text{WAL}(m, t) = \text{WAL}(n \oplus m, t) \qquad (2.15)$$

as may be seen from the following.
From Eq. (1.15) we have

$$\text{WAL}(n, t) = \prod_{r=0}^{p-1} (-1)^{n_{p-1-r}(t_r + t_{r+1})} \qquad (2.16)$$

and

$$\text{WAL}(m, t) = \prod_{r=0}^{p-1} (-1)^{m_{p-1-r}(t_r + t_{r+1})} \qquad (2.17)$$

where n, m and t are expressed in p binary terms and $N = 2^p$.
The product of the two functions will be

$$\begin{aligned}
\text{WAL}(n, t)\,\text{WAL}(m, t) &= \prod_{r=0}^{p-1} (-1)^{(n_{p-1-r} + m_{p-1-r})(t_r + t_{r+1})} \\
&= \prod_{r=0}^{p-1} (-1)^{(n \oplus m)_{p-1-r}(t_r + t_{r+1})} \\
&= \text{WAL}[(n \oplus m), t] \qquad (2.18)
\end{aligned}$$

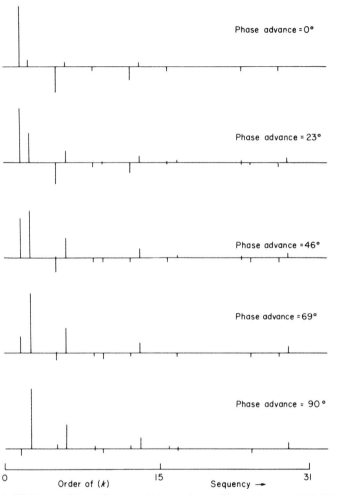

FIG. 2.1. Walsh transforms of a sinusoidal waveform with circular time shift ($N = 32$).

since the addition of binary terms of the same index must be carried out by modulo-2 addition.

By using Eq. (1.22) the following set of relationships can also be obtained:

$$\begin{aligned}
&\text{CAL}(k, t)\, \text{CAL}(p, t) = \text{CAL}(k \oplus p, t) \\
&\text{SAL}(k, t)\, \text{CAL}(p, t) = \text{SAL}(p \oplus (k-1) + 1, t) \\
&\text{CAL}(k, t)\, \text{SAL}(p, t) = \text{SAL}(k \oplus (p-1) + 1, t) \quad (2.19)\\
&\text{SAL}(k, t)\, \text{SAL}(p, t) = \text{CAL}((k-1) \oplus (p-1), t) \\
&\text{CAL}(0, t) = \text{WAL}(0, t)
\end{aligned}$$

2.2 The Discrete Walsh Transform

which correspond to the set of circular function relationships

$$2\cos kt \cos pt = \cos(k-p)t + \cos(k+p)t$$
$$2\sin kt \cos pt = \sin(k-p)t + \sin(k+p)t$$
$$2\cos kt \sin pt = -\sin(k-p)t + \sin(k+p)t$$
$$2\sin kt \sin pt = \cos(k-p)t - \cos(k+p)t$$

(2.20)

An important difference between these two sets of relationships lies, however, in the absence of a shift theorem in the Walsh case. Whilst the shift theorem enables the product of two Fourier transforms to be inversely transformed to obtain a convolution of the two original time series, a similar result is not obtained with the Walsh transform. This is considered further in Chapter 3, where convolution and correlation using Walsh functions are considered.

If the addition relationship of Eq. (2.15) is combined with the symmetry relationship given in Eq. (1.23), then we have the interesting result

$$\text{WAL}(k \oplus p, t) = \text{WAL}(k, t)\, \text{WAL}(p, t)$$
$$= \text{WAL}(t, k)\, \text{WAL}(t, p)$$
$$= \text{WAL}(t, k \oplus p) \quad (2.21)$$

This can provide the means for generation of a series of Walsh functions from the symmetry relationship alone [7].

The characteristics of the discrete Walsh series and its transform discussed earlier are compared with the discrete Fourier series and transform and summarised in Table 2.2. The conversion between the two transforms will be described next.

2.2.4 Conversion between discrete Walsh and Fourier transformation

From Eqs. (2.6) and (2.9) we can express a sampled time series x_i of size N in terms of its Fourier transform series X_f or its Walsh transform series X_n, viz.,

$$x_i = \sum_{f=0}^{N-1} X_f \exp(j2\pi i f / N) = \sum_{n=0}^{N-1} X_n\, \text{WAL}(n, i) \quad (2.22)$$

where $f, i, n = 0, 1, \ldots, N-1$.

Hence conversion from Fourier to Walsh transformation is

$$X_n = \frac{1}{N} \sum_{i=0}^{N-1} x_i\, \text{WAL}(n, i)$$
$$= \frac{1}{N} \sum_{f=0}^{N-1} X_f \left(\sum_{i=0}^{N-1} \text{WAL}(n, i) \exp(j2\pi i f / N) \right) \quad (2.23)$$

TABLE 2.2. Summary of characteristics for Walsh and Fourier discrete series of length N.

	Walsh			Fourier
Time index	i		i	$j = \sqrt{-1}$
Sequency index	k		n	
Walsh order index	n	$i, n = 0, 1, 2, \ldots, N-1$		
CAL function	$\text{CAL}(k, i) = \text{WAL}(2k, i)$		Cos function	$\cos\left(\dfrac{2\pi n i}{N}\right)$
SAL function	$\text{SAL}(k, i) = \text{WAL}(2k-1, i)$		Sin function	$\sin\left(\dfrac{2\pi n i}{N}\right)$
Walsh function	$\text{WAL}(n, i) = \text{WAL}(i, n)$		Complex function	$\exp\left(j\dfrac{2\pi n i}{N}\right)$ $= \cos\left(\dfrac{2\pi n i}{N}\right) + j\sin\left(\dfrac{2\pi n i}{N}\right)$
Time series	$x_i = a_0\,\text{WAL}(0, i) + a_1\,\text{WAL}(1, i)$ $+ \cdots + a_{N-1}\,\text{WAL}(N-1, i)$		Time series	$x_i = A_0 + A_1 \exp\left(\dfrac{j2\pi i}{N}\right)$ $+ A_2 \exp\left(\dfrac{j4\pi i}{N}\right) + \cdots + A_N \exp(j2\pi i)$

Sequency coefficients	$a_0, a_1, \ldots, a_{N-1}$	Complex frequency coefficients	$A_0, A_1, \ldots, A_N$
Walsh series	$x_i = a_0 \, \text{WAL}(0, 1)$ $\quad + \sum_{m=1}^{N/2} \sum_{p=1}^{N/2-1} a_m \, \text{SAL}(m, i) + b_p \, \text{CAL}(p, i)$	Fourier series	$x_i = \dfrac{a_0}{2} + \sum_{n=1}^{N/2} \left(a_n \cos\left(\dfrac{2\pi n i}{N}\right) \right.$ $\left. + b_n \sin\left(\dfrac{2\pi n i}{N}\right) \right)$
Walsh transform	$X_n = \dfrac{1}{N} \sum_{i=0}^{N-1} x_i \, \text{WAL}(n, i)$	Fourier transform	$X_n = \dfrac{1}{N} \sum_{i=0}^{N-1} x_i \exp\left(-j\dfrac{2\pi n i}{N}\right)$
Walsh inverse transform	$x_i = \sum_{n=0}^{N-1} X_n \, \text{WAL}(n, i)$	Fourier inverse transform	$x_i = \sum_{i=0}^{N-1} X_n \exp\left(j\dfrac{2\pi n i}{N}\right)$
Walsh CAL transform	$X_c(k) = \dfrac{1}{N} \sum_{i=0}^{N-1} x_i \, \text{CAL}(k, i)$	Fourier cosine transform	$X_c(f) = \dfrac{1}{N} \sum_{i=0}^{N-1} x_i \cos\left(\dfrac{2\pi n i}{N}\right)$
	$f, k = 1, 2, \ldots, N/2$		
Walsh SAL transform	$X_s(k) = \dfrac{1}{N} \sum_{i=0}^{N-1} x_i \, \text{SAL}(k, i)$	Fourier sine transform	$X_s(f) = \dfrac{1}{N} \sum_{i=0}^{N-1} x_i \sin\left(\dfrac{2\pi n i}{N}\right)$

and conversion from Walsh to Fourier transformation is

$$X_f = \frac{1}{N} \sum_{i=0}^{N-1} x_i \exp(-j2\pi i f/N)$$

$$= \frac{1}{N} \sum_{n=0}^{N-1} X_n \left(\sum_{i=0}^{N-1} \text{WAL}(n, i) \exp(-j2\pi i f/N) \right) \quad (2.24)$$

The limitation here is, of course, that N be sufficiently large so that an accurate representation of the sampled time series is possible from a limited number of terms.

There are some advantages in deriving the discrete Fourier transform from a version of the fast Walsh transform. This will be discussed in Section 2.7.

2.3 Fast Walsh transform algorithms

The matrix representation of a discrete Walsh function in natural order was given in Subsection 1.3.4. If we define this as $\mathbf{H}_N$, then the discrete Walsh transform, corresponding to Eq. (2.5), may be expressed as

$$\mathbf{X}_N = \mathbf{H}_N x_i \quad (2.25)$$

where x_i is a column vector of order N representing the sampled values of an input signal.

The transformation is carried out by multiplying the input data string x_i with elements of the Hadamard matrix $\mathbf{H}_N$ and proceeding down each column in turn and summing each set of column products to obtain the transformed output data string $\mathbf{X}_N$. Since the products represent multiplication by $+1$ or -1, this procedure will be seen to require $N(N-1)$ additions or subtractions. However, as shown by Good [1] and others it is possible to simplify the task through a *matrix factorisation technique* which reveals a large amount of redundancy in the repeated calculations implied in Eq. (2.25).

Good's technique can be used to factor matrices derived from Kronecker products into p factors of order $N = 2^p$, where N is the order of the original matrix and p the number of matrices in the Kronecker product. Where the original matrix is derived from the Kronecker product of p identical matrices, as is the case for the derivation of the Walsh matrix in natural order described in Subsection 1.3.4, all the p factors are identical. This is the simplest case, and for $N = 8$ the equivalence can be expressed as

$$\mathbf{H}_N = \begin{bmatrix} 1 & 1 & 1 & 1 & 1 & 1 & 1 & 1 \\ 1 & -1 & 1 & -1 & 1 & -1 & 1 & -1 \\ 1 & 1 & -1 & -1 & 1 & 1 & -1 & -1 \\ 1 & -1 & -1 & 1 & 1 & -1 & -1 & 1 \\ 1 & 1 & 1 & 1 & -1 & -1 & -1 & -1 \\ 1 & -1 & 1 & -1 & -1 & 1 & -1 & 1 \\ 1 & 1 & -1 & -1 & -1 & -1 & 1 & 1 \\ 1 & -1 & -1 & 1 & -1 & 1 & 1 & -1 \end{bmatrix}$$

$$= \begin{bmatrix} 1 & 1 & 0 & 0 & 0 & 0 & 0 & 0 \\ 0 & 0 & 1 & 1 & 0 & 0 & 0 & 0 \\ 0 & 0 & 0 & 0 & 1 & 1 & 0 & 0 \\ 0 & 0 & 0 & 0 & 0 & 0 & 1 & 1 \\ 1 & -1 & 0 & 0 & 0 & 0 & 0 & 0 \\ 0 & 0 & 1 & -1 & 0 & 0 & 0 & 0 \\ 0 & 0 & 0 & 0 & 1 & -1 & 0 & 0 \\ 0 & 0 & 0 & 0 & 0 & 0 & 1 & -1 \end{bmatrix}^3 \quad (2.26)$$

It is the presence of so many zero components in this factored form for $\mathbf{H}_N$ that leads directly to a considerable reduction in the product calculations required in the calculation of the discrete transform from (in this case) $N(N-1)$ to $N \log_2 N$ additions and subtractions.

There are several ways in which this redundancy can be exploited to obtain a fast transform algorithm. The method chosen is dependent on the characteristics of the transformation required, and although matrix methods are widely used, other representations are possible. For the purpose of illustration, a continued product representation of the Walsh function is used in the derivation of a sequency-ordered fast transform which follows.

2.3.1 A sequency-ordered fast transform

A derivation of a fast Walsh transform (FWT) algorithm having sequency-ordered coefficients may be obtained from the continued product representation given by Pratt et al. [8]. This was defined in Eq. (1.15) for a series of $N = 2^p$ terms as

$$\text{WAL}(n, i) = \prod_{r=0}^{p-1} (-1)^{n_{p-1-r}(i_r + i_{r+1})}$$

$$i, n = 0, 1, 2, \ldots, N-1, \quad r = 0, 1, 2, \ldots, p \quad (2.27)$$

Here i, n are expressed in terms of their binary digits, i_r and n_r, e.g.,

$$i = (i_{p-1}, i_{p-2}, \ldots, i_1, i_0)_2 \qquad (2.28)$$

The algorithm is developed by substitution of Eq. (2.27) into Eq. (2.5) and factorising the calculation into p separate stages. Carrying out this substitution we otain a product–sum expression for the discrete Walsh transform as

$$X_n = \sum_{i=0}^{N-1} x_i \, \text{WAL}(n, i) = \prod_{r=0}^{p-1} \sum_{i_r=0}^{1} (-1)^{n_{p-1-r}(i_r+i_{r+1})} x_{(i_{p-1},\ldots,i_0)} \qquad (2.29)$$

(neglecting scaling by N). Here x_i is expressed as $x_{(i_{p-1},\ldots,i_0)}$ and X_n is expressed as $X_{(n_{p-1},\ldots,n_0)}$, where i_{p_r} and n_p are the binary bits of i and n with $r = 0, 1, 2, \ldots, p$.

The calculation of the FWT is carried out in a series of stages, one stage for each power 2 for N. The first calculation stage is to derive a partial transformation series, A_n from the input series x_i by placing $r = 0$ in Eq. (2.29), giving

$$A(n_{p-1}, i_{p-1}, \ldots, i_1) = \sum_{i_0=0}^{1} (-1)^{n_{p-1}(i_0+i_1)} x_{(i_{p-1},\ldots,i_0)} \qquad (2.30)$$

To see how this is calculated the case of $N = 8$ can be studied. We take the first adjacent pair of data samples and look at the i_1 bit in x_i. Four values of

$$(-1)^{n_{p-1}(i_0+i_1)} \quad \text{for} \quad i_0 = 0 \text{ or } 1, \quad n_{p-1} = 0 \text{ or } 1$$

are obtained. These indicate whether to add or subtract the two adjacent values of x_i and x_{i+1} (where i and $i+1$ are decimal values). From these sums and differences the two intermediate transformed values are obtained as

$$x_i + x_{i+1}$$

and either

$$x_i - x_{i+1} \quad \text{or} \quad x_{i+1} - x_i$$

depending on the sign for $(-1)^{n_{p-1}(i_0+i_1)}$ worked out earlier. This process is continued for the remaining consecutive pairs of x_i coefficients.

Further stages of calculation proceed serially by using the results of the preceding stage as input for the next stage. This may be expressed in the general expression for the intermediate transformation as

$$A_r(n_{p-1}, n_{p-2}, \ldots, n_{p-r}; i_{p-1}, i_{p-2}, \ldots, i_r)$$
$$= \sum_{i_{r-1}=0}^{1} (-1)^{n_{p-r}(i_r+i_{r-1})} A_{r-1}(n_{p-1}, \ldots, n_{p-r+1}, i_{p-1}, \ldots, i_{r-1}) \qquad (2.31)$$

with the values of A_1 tp A_p retained in temporary storage during the course of the calculation.

2.3 Fast Walsh Transform Algorithms

Finally we need to normalise the result through division by N, viz.,

$$X_{(n_{p-1},n_{p-2},...,n_0)} = (1/N)A_{p(n_{p-1},n_{p-2},...,n_0)} \tag{2.32}$$

Thus it is seen that the complete transform may be obtained in $N \log_2 N$ addition and subtraction operations rather than in N^2 operations demanded by the direct method of calculation for Eq. (2.5).

2.3.2 Signal flow diagrams

The mathematical procedures for transform algorithms have been adequately described in the literature and will not be pursued further here. Instead the characteristic features of these various algorithms will be considered in order that their application to signal processing, control and other problems may be more fully appreciated.

A common feature of all the fast discrete transformations described is the possibility of expressing their algorithmic construction in the form of a signal flow diagram. From this, together with the symmetry features of the particular derivation, it is a fairly straightforward task to write a computer program in one of the common high-level languages, Fortran, Algol, Pascal, etc., or to construct hardware logic to carry out the particular transformation required.

In what follows the characteristics of a number of discrete sequency transformations will be described and use made of signal flow diagrams to explain their operation and how they compare with each other.

The principle of a signal flow diagram is quite simple. The diagram consists of a series of nodes, each representing a variable which is itself expressed as the sum of other variables originating from the left of the diagram, with the nodes connected together by means of straight lines. The weighting of these additions is shown by a number appearing at the side of an indicating arrow on these connecting lines. Thus from Fig. 2.2, the variable $A7$ is derived from variables originating at nodes $B3$ and $C3$, with the latter weighted by 2, so that we can write

$$A7 = B3 + 2 \cdot C3 \tag{2.33}$$

Since the computation of many sequency transforms consists of either repeated addition or subtraction of pairs of terms, the convention adopted in this book is to indicate a term to be subtracted by a dashed connecting line and an added term by a full line.

Figure 2.3 shows the flow diagram for the sequency-ordered fast Walsh transform ($N = 8$), the derivation for which we considered in the previous section. This figure shows two flow diagram features that are found in almost all of the fast transformation algorithms.

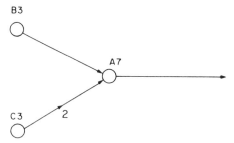

Fig. 2.2. A signal flow diagram.

The first is the concept of taking a pair of adjacent terms and either adding or subtracting the pair to produce a new term for a subsequent stage of the algorithm. A similar feature, which we shall discuss later, occurs when the pair are not adjacent to each other but separated by $N/2$ samples. The second feature, also contained in Fig. 2.3, occurs when a series of identical self-contained operations are carried out on two values only. This is reproduced in Fig. 2.4a and has been referred to as a *butterfly* diagram. Here the output value D_1 is obtained from a linear combination of C_1 and C_2, whilst D_2 is

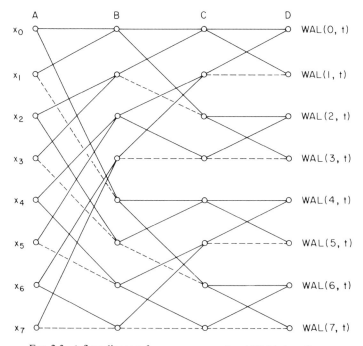

Fig. 2.3. A flow diagram for a sequency-ordered Walsh transform.

2.3 Fast Walsh Transform Algorithms

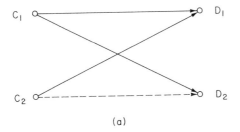

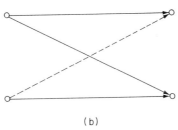

FIG. 2.4. (a) A normal butterfly, (b) a reversed butterfly.

obtained from the combination C_1 and C_2, with the latter modified by a sign inversion.

These butterflies have the inherent advantage that each of the transformation steps, once a pair of data locations have been read, can be overwritten by the calculated pair of output values. This is the key to *in-place* algorithms in which memory storage for intermediate-stage calculations is not needed since the calculated values can be placed back into memory locations occupied previously by the initial data values.

A similar feature, often referred to as a *reversed butterfly* (illustrated in Fig. 2.4b), is also found.

2.3.3 Conversion from matrix to signal flow diagram

Matrix methods for the derivation of sequency transformation algorithms are widely used and given a solution in terms of a series of matrix products (for example, Eq. (2.26)). It is a fairly simple matter to obtain the equivalent signal flow diagram and hence derive a computational algorithm. This is shown in Fig. 2.5 for the $N = 8$ matrix expressed by the right-hand side of Eq. (2.26). Here the columns have been labelled 'input series' $x_0, x_1 \ldots,$

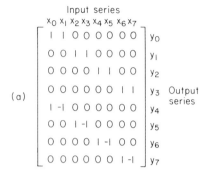

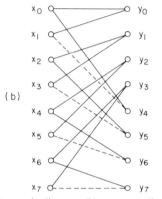

FIG. 2.5. (a) A matrix diagram, (b) corresponding signal flow diagram.

x_7 and the rows as the 'output series' $y_0, y_1, \ldots, y_7$. The single-stage flow diagram corresponding to this matrix is given in Fig. 2.5b, where $x_0, x_1, \ldots, x_7$ represent the input values to the stage and $y_0, y_1, \ldots, y_7$ the combined output values.

To select the pair of input terms to be combined to form a single output term it is necessary to scan the columns of the matrix to find a pair of values other than zero along the output series row. For example, row y_0 contains entries 1 and 1 for columns x_0 and x_1, enabling the appropriate lines connecting x_0 and x_1 to y_0 to be drawn in the flow diagram. Where this scan along a row reveals two pairs of opposite signs e.g., 1 and -1 found in columns x_4 and x_5 in the output series row y_6, the line connecting x_5 to y_6 is shown dotted, indicating a subtraction $x_4 - x_5$. If the matrix value at the conjugation an x and y series has a value other than $+1$ or -1, then this represents a multiplying factor, such as the 2 shown in Fig. 2.2.

2.3.4 Ordering in fast Walsh transform algorithms

We saw earlier that, due to the mathematical nature of the Walsh function series, a single universal ordering for the Walsh function does not exist. In contrast to the complex exponential functions which are always ordered by their arithmetic value, i.e., frequency, we need to consider at least three different orderings of the Walsh function and hence at least three kinds of transform algorithms [9].

The simplest in terms of computer organisation are those resulting in natural-ordered transformed values and derived from Kronecker products of the Hadamard matrix (Subsection 1.3.4). This is essentially the way the FFT was originally obtained. It is possible to derive a suitable computer program by modification of some Cooley–Tukey (C–T) fast transform algorithms [2]. The object is to reduce the trigonometric values used in the program to unity and to remove the complex part of the operation since the Walsh transform is a real one. Often a dummy parameter is included in the calculation for the exponent (Eq. (2.9)) which is made $+1$ or -1, depending on whether a direct or inverse Fourier transform is required. By making this parameter 0, the exponentials are reduced to ± 1 and a dyadic- (bit-reversed natural) ordered FWT results.

The type of modified C–T algorithm resulting is shown in Fig. 2.6, which is based on the *decimation in frequency* form of the FFT and has been described by Welchel and Guinn [4].

An implementation of this, derived from the mathematical description of the Walsh transform, is given by Shanks [10]. In this implementation, data are accepted in normal order but the transformed output is in dyadic order. Since the algorithm is seen to be constructed entirely of butterfly subsections in which each pair of nodes in a butterfly only affect a pair of nodes immediately to the right, it is an *in-place* algorithm. That is, no additional register or memory storage is required. After each new pair of results is computed, they can be located in the register in place of the initial pair of values, which is no longer needed.

Maintaining the input nodes x_0 to x_7 in their original places in Fig. 2.6 and rearranging the output nodes designated PAL(0, t), PAL(1, t), etc., in the order appropriate to Hadamard-ordered output and carrying all the connecting lines with them results in the other familiar form of the C–T algorithm in which the input (and not the output) is in bit-reversed form. The corresponding flow diagram for Hadamard output ordering is given in Fig. 2.7. This is also an in-place algorithm. It has been pointed out by Kunt [11] that this form allows the simple derivation of a phase-invariant transform, the R transform, which will be discussed later.

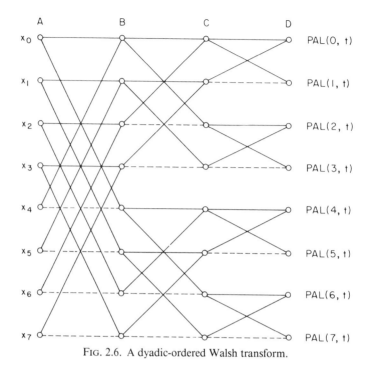

FIG. 2.6. A dyadic-ordered Walsh transform.

The Walsh transform resulting from the Kronecker multiplication of the Hadamard matrix is often referred to as the Walsh–Hadamard transform (WHT) or simply the Hadamard transform (both terms will be found in this book). A difficulty with the WHT is that to obtain a sequency-ordered output, which is desirable for many applications, it is necessary to carry out a bit reversal on the input or output terms together with a Gray code reordering of the output.

A method of avoiding this Gray code reordering was first suggested by Manz [12]. He replaced some of the normal butterflies of the C–T-derived algorithm (Fig. 2.6) by reversed ones (Fig. 2.4b) and further stated a series of rules by which the blocks of butterflies for reversal may be selected. With reference to Fig. 2.6 with $N = 2^p = 8$, the following rules apply:

Rule 1 Lines originating from the input nodes A are *never* reversed.

Rule 2 The next set of nodes between B and C are defined as having a number of 'blocks' of butterflies. A block denotes neighbours above and below. Thus two such blocks are present between B and C and four between C and D. In general, commencing with $2^0 = 1$ blocks between A and B, $2^1 = 2$ blocks between C and D, etc., we would expect to find $2^{p-1} = N/2$ blocks between the final pair of columns of nodes.

2.3 Fast Walsh Transform Algorithms

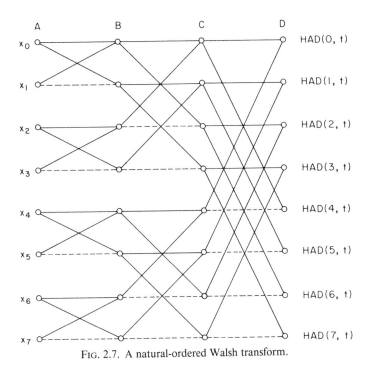

FIG. 2.7. A natural-ordered Walsh transform.

Rule 3 Reversals occur in blocks. At each column of nodes after the first the bottom block is reversed. Alternate blocks are then reversed, with the top block always remaining unreversed.

The flow diagram for Manz's algorithm is shown in Fig. 2.8 for $N = 8$. This gives output values in sequency order, but the input is in bit-reversed order. Like the other algorithms derived from the C–T FFT, this is an in-place algorithm and remains its own inverse so that a separate inverse Walsh transform is not required.

A similar algorithm derived from the alternative C–T flow diagram (Fig. 2.7) accepts data in sequential order but gives the transformed coefficients in bit-reversed sequency order (Fig. 2.9). This is due to Larsen [13], who states that this order has advantages in power spectral estimation since the spectral components can also be averaged in bit-reversed order. (See Chapter 3.)

For those applications requiring a double transformation (e.g., a transformation followed by some processing and an inverse transformation), the use of the Larsen transform followed by the Manz transform could eliminate the bit inversion stage completely. In this sense the two transforms could be regarded as complementary to each other.

The similarities in mathematical form of the Fourier, Walsh and other

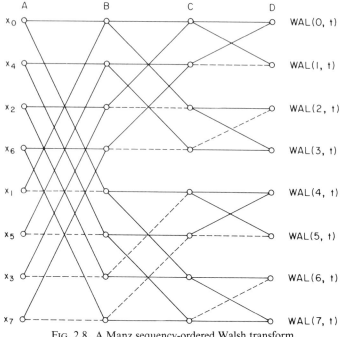

FIG. 2.8. A Manz sequency-ordered Walsh transform.

transforms derived from the C–T algorithm have suggested the development of a generalised transform which can express any of these. Caspari [14] has developed such a transform, which has led to the implementation of a fast transform algorithm by Ahmed and Rao [15] which includes the Walsh and Fourier transforms as special cases. These implementations are accomplished through a Kronecker product of a set of sparse matrices, as discussed earlier, or by matrix factorisation. The computational algorithm describes an identical sequence of operations for all the possible transforms, with only the multiplying factors changing. This is shown in terms of a signal flow diagram in Fig. 2.10 for $N = 8$. The multiplying factors for the Walsh and Fourier transforms are shown in Table 2.3.

The general properties of orthogonal matrices, from which many of the fast transformations are derived, include the systematic alteration of signs within the matrix and interchange of any row or column with any other row or column without impairing the orthogonality of the matrix. Because of these features we find that a large number of variations of possible orthogonal transforms are feasible. It has been shown, for example, by Madych [16], that for $N = 2^p$ there are $p = \log_2 N$ possible discrete orthogonal transforms using the generalised transform algorithm, although few of these appear to have useful applications.

2.3 Fast Walsh Transform Algorithms

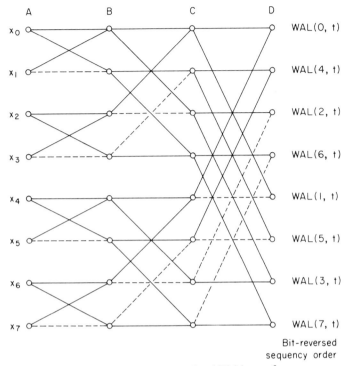

FIG. 2.9. A Larsen sequency-ordered Walsh transform.

Recently the application of the ideas of the *perfect shuffle* (discussed later in Subsection 4.3.1) and Kronecker products for the factorisation of orthogonal transform matrices has led to a unified structure for the development of fast algorithms. This is proving valuable as a powerful tool for investigation of recursive properties, reordering techniques and fast algorithms, some of which are described here and in Chapter 4 [16–18, 30].

The early work of Good [1], which forms the basis of the type of *FFT-derived* transforms we have been discussing, also contains a quite different form of matrix factorisation which has been used for several sequency transform generations. One of these is the sequency-ordered Walsh transform algorithm shown in Fig. 2.3 and described earlier in Subsection 2.3.1. This transform is characteristic of all sequency-ordered algorithms which are *not* able to carry out all of their intermediate calculations in place. In addition to the normal butterflies this flow diagram also includes a number of paired terms in which the algorithm performs two operations on each of the N vector elements to be processed (an addition and a subtraction). The products must be held in a buffer location. Each succeeding operation product is stored in a buffer which, in computing terms, operates as a push-up stack;

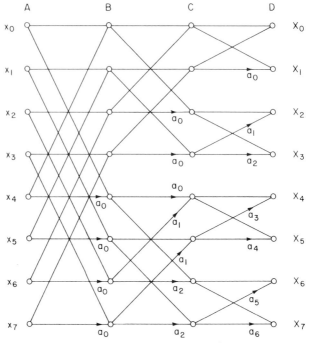

FIG. 2.10. The generalised transform.

i.e., the new entering product causes the oldest product to be removed from the buffer and the new product written in a location occupied by the last vector element to be completely processed.

A complementary form of this algorithm, due to Fontaine [19], is shown in Fig. 2.11 and also requires N additional storage locations for the intermediate calculations. This formulation corresponds exactly with the sum and difference method used by Blackman and Tukey [20] in their well-known method for the estimation of a power spectra. This applies to the intermediate steps in the transform calculation as well since they are in sequency order. Hence the method may be used directly for time-dependent spectral analysis if the input data form a sampled time series. Brown [21] arrives at

TABLE 2.3. Multipliers for the generalised transform signal flow graph ($N = 8$).[a]

	a_0	a_1	a_2	a_3	a_4	a_5	a_6
Walsh	-1	1	-1	1	-1	1	-1
Fourier	W^4	W^2	W^6	W^1	W^5	W^3	W^7

[a] $W = \exp(-j2\pi i f/8)$.

2.3 Fast Walsh Transform Algorithms

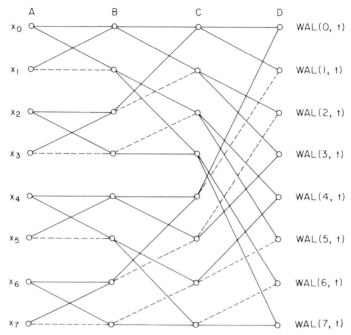

FIG. 2.11. A Fontaine sequency-ordered Walsh transform.

the same algorithm through factorising the Hadamard matrix into N sparse matrices which, although different from each other, are similar to those given in Eq. (2.26).

A number of other sequency-ordered Walsh routines have been described elsewhere [22, 23]. A derivation from the difference equation noted in Subsection 1.3.4 is due to Harmuth [22] and is given in Fig. 2.12. Apart from the lack of symmetry in the flow diagram which would make programing difficult, it assumes a Walsh function series based on the interval $(-\frac{1}{2}, 0, \frac{1}{2})$, which may not be desirable.

Ulman's algorithm [23], shown in Fig. 2.13, has the advantage that, like others in which pairs of input vectors are separated by $N/2$ places, it is possible to derive the R transform by a simple change in one computer instruction enabling absolute values to be used for the results of each intermediate computation which is indicated as a subtraction. (The cyclic invariant R transform is considered in Section 2.6.)

Another alternative route to the sequency-ordered transform is to employ the more efficient in-place algorithm and carry out bit inversion by means of a separate routine prior to or after the transform. Several efficient routines to

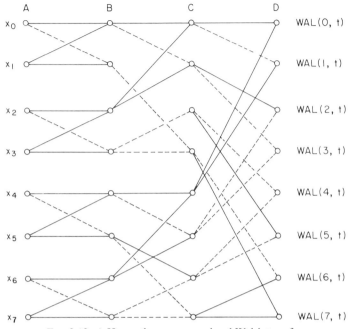

FIG. 2.12. A Harmuth sequency-ordered Walsh transform.

do this are described in the literature, and the combination is often faster than where the transform algorithm itself includes the bit-reversal routine [11, 24].

The original matrix decomposition (due to Good) upon which the algorithms of Pratt and Brown are based (Figs. 2.3 and 2.11) contains identical nodal column stages. This feature was discussed earlier and the matrix equation given in Eq. (2.26). A flow diagram for a Walsh transform algorithm based directly on this is given in Fig. 2.14. This gives an output in natural order and will require auxiliary storage for intermediate stage results. Note that only paired terms are involved and that no self-contained butterfly loops are employed.

This repeated type of flow diagram is known as a *constant-geometry* structure. There are obvious advantages in hardware construction [25, 26] of the transform algorithm, particularly where it is desired to construct a programmable logic array of the device or other implementation using integrated circuits [27]. (See Chapter 4.)

The recursive structure of this algorithm can be expressed as follows. Referring to Fig. 2.14 and designating the coefficient value at each node position as $y_i(n)$, where $i = 0, 1, 2, 3$ and with n is the position down each

2.3 Fast Walsh Transform Algorithms

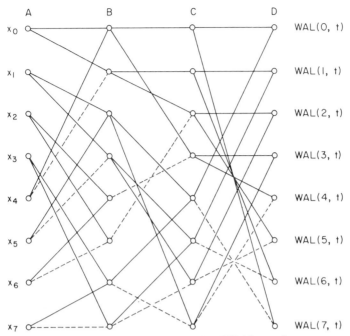

FIG. 2.13. An Ulman sequency-ordered Walsh transform.

column of nodes $n = 1, 2, \ldots, 8 \ (= N = 2^p)$, the input, intermediate and transformed values are, respectively,

$$y_0(n) = x(n)$$

$$y_{i+1}(n) = \begin{cases} y_i(2n-1) + y_i(2n) & \text{for } 0 < n \leq N/2 \\ y_i(2n-N-1) - y_i(2n-N) & \text{for } n/2 < n \leq N \end{cases} \quad (2.34)$$

These results can be extended easily into higher dimensions and programmed accordingly. The program will require a working area of at least $N/2$ locations as well as N places for the data.

An earlier constant-geometry configuration is due to Geadah and Corinthios [28]. This is developed from the Kronecker matrix expansion which is pre-multiplied by a permutation matrix performing a bit reversal so that a dyadic-ordered output is obtained. The restructured matrix expansion results in a multi-stage algorithm in which p identical stages are used, each consisting of $N/2$ normal butterflies (Fig. 2.15). These stages are linked by means of a simple reordering routine. Again the benefit here is in logical hardware construction for the transform.

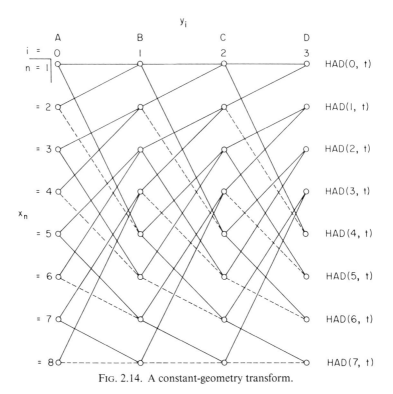

FIG. 2.14. A constant-geometry transform.

2.3.5 CAL–SAL transformation

A further ordering of the Walsh function series is the CAL–SAL ordering introduced in Subsection 1.3.2. Fast transform algorithms have been described for this by Rao [29] and Kunt [30]. Both CAL and SAL are derived from the factorisation of the equivalent Hadamard matrix obtained by periodic sampling of the set of functions shown in Fig. 1.11. In the case of Rao's algorithm, the products of the p sparse matrices so formed lead to the flow diagram given in Fig. 2.16. This requires $2N$ storage locations and gives the CAL–SAL ordering directly.

The second algorithm, due to Kunt, is the in-place algorithm shown in Fig. 2.17, which is similar to the conventional transform obtained from the C–T FFT (Fig. 2.7). This algorithm differs from the C–T algorithm, however, by the inclusion of reversed butterflies to ensure the correct ordering. This is not obtained directly, but through an additional bit-reversing sequence and an inverting sequence (first in . . . last out) for the second half of the coefficients (not shown), since these are otherwise obtained in increasing order rather than in the decreasing SAL order required by the definition of the CAL–SAL transform. It may be noted that if the application for this is

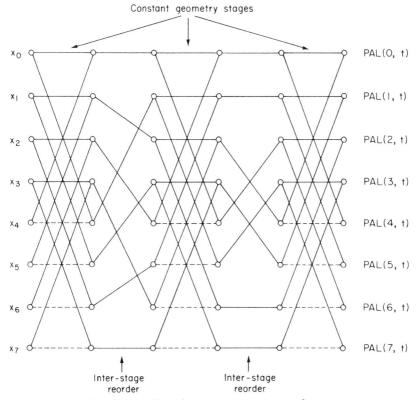

FIG. 2.15. A Geadah constant-geometry transform.

only concerned with determining whether the applied signal has even or odd symmetry, then this bit-reversing routine can be omitted.

A summary of the characteristics of the principal fast Walsh transforms we have been considering is given in Table 2.4.

2.4 The discrete Haar transform

From Eq. (1.72) and (1.73) the discrete Haar transform and its inverse can be stated as

$$\mathbf{X}_n = \frac{1}{N} \sum_{i=0}^{N-1} \mathbf{x}_i \, \mathrm{HAR}(n, i/N) \qquad (2.35)$$

$$\mathbf{x}_i = \sum_{n=0}^{N-1} \mathbf{X}_n \, \mathrm{HAR}(n, i/N) \qquad (2.36)$$

where $i, n = 0, 1, \ldots, N-1$. Written in matrix form, Eqs. (2.35) and

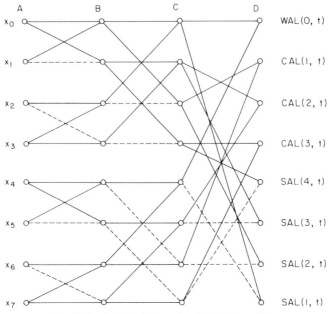

Fig. 2.16. A CAL–SAL-ordered Walsh transform.

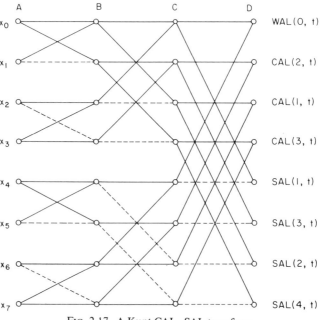

Fig. 2.17. A Kunt CAL–SAL transform.

TABLE 2.4. Fast Walsh transform algorithms.

Figure	Author(s)	Order of input data	Order of output data	In place	Geometry	Derivation	Features
2.6	Shanks	Sequential	Dyadic	Yes	Normal butterflies	Modified C–T FFT	Requires bit inversion and Gray code conversion to obtain a sequency output
2.7	Kunt	Bit reversed	Natural	Yes	Normal butterflies	Modified C–T FFT	Allows simple conversion to R transform
2.8	Manz	Bit reversed	Sequency	Yes	Normal and reversed butterflies	Modified C–T FFT	Avoids Gray code conversion of output
2.9	Larson	Sequential	Bit reversed, sequency	Yes	Normal and reversed butterflies	Modified C–T FFT	Is complementary to Manz algorithm; use of Manz and Larson together avoids bit reversal in a double transformation
2.10	Caspari	Sequential	Bit reversed, natural (dyadic)	Yes	Normal butterflies	Kronecker product or modified C–T FFT	Involves Fourier, Walsh or complex Walsh obtained by choice of multiplication constants
2.3	Pratt	Sequential	Sequency	No	Paired terms and normal butterflies	Continued product	Is slower than the in-place algorithms.
2.11	Fontaine	Sequential	Sequency	No	Normal and reversed butterflies and paired terms	Kronecker product	Can be used for time-dependent spectral analysis
2.12	Harmuth	Sequential	Sequency	No	Reversed butterflies and paired terms	Difference equation	Assumes an open interval of $\pm \frac{1}{2}t$ for the WF
2.13	Ulman	Sequential	Sequency	No	Paired terms	Difference equation	Allows simple conversion to R transform
2.14	Carl and Swartwood	Sequential	Natural	No	Constant with paired terms	Kronecker product	Is suitable for hardware implementation
2.15	Geadah	Sequential	Dyadic	No	Constant with normal butterflies and paired terms	Kronecker product	Was developed for hardware implementation and requires interstage reordering
2.16	Rao	Sequential	CAL–SAL	No	Normal and reversed butterflies and paired terms	Kronecker product	Is useful when transforming an even or odd series
2.17	Kunt	Sequential	Bit reversed, CAL–SAL	Yes	Normal and reversed butterflies	Kronecker product	Requires bit inversion and inversion sequences to obtain CAL–SAL ordering

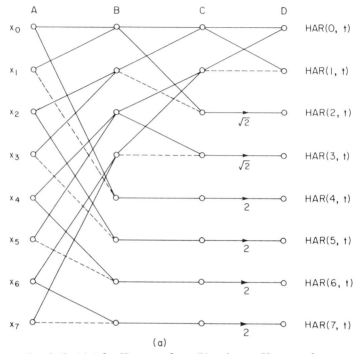

FIG. 2.18. (a) A fast Haar transform, (b) an inverse Haar transform.

(2.36) become

$$\mathbf{X} = \frac{1}{N} \cdot \mathbf{Ha} \cdot \mathbf{x} \tag{2.37}$$

and

$$\mathbf{x} = \mathbf{Ha}^{-1} \cdot \mathbf{X} \tag{2.38}$$

where **Ha** and **Ha**$^{-1}$ are the direct and transposed Haar function matrices. Unlike the Walsh transform, the matrix is not symmetric, so that separate transform operations are required for transformation and inverse transformation.

2.4.1 Fast Haar transform algorithms

If the transformation given in Eq. (2.35) is carried out directly, then N^2 additions will become necessary. This can be reduced to pN, where $N = 2^p$ if only the non-zero values are considered.

2.4 The Discrete Haar Transform

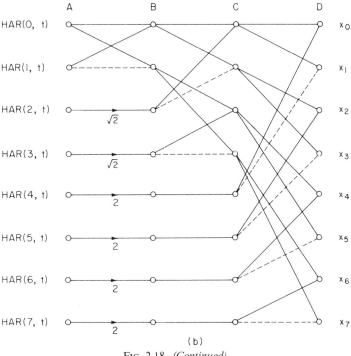

FIG. 2.18. *(Continued)*

A considerable improvement in computational efficiency is obtained if a factorisation algorithm similar to that used for the fast Fourier and Walsh transforms is employed. A flow diagram for an eight-point fast Haar transform due to Rejchrt [31] is shown in Fig. 2.18a. Multiplication of the sum/differences by 1 or $\sqrt{2}$ is indicated in the diagram. It will be seen that at each step in the calculation (other than with the first) half the points require no further calculation. (The multiplication can all be delayed until the transformation is complete.) Thus, the total number of additions or subtractions is

$$N + (N/2) + (N/4) + \cdots + 2 = 2(N-1) \qquad (2.39)$$

Transformation time for this algorithm is therefore linearly proportional to the number of terms N, in contrast to the fast Walsh or Fourier transform, where it is proportional to $N \log_2 N$. Since the matrix for the Haar transform is not symmetric, a separate inverse transform is required. A flow diagram for this is shown in Fig. 2.18b.

As will be apparent from an inspection of the flow diagrams, additional storage will be required to hold the intermediate stage calculations. This computational overhead can be avoided with the in-place algorithm de-

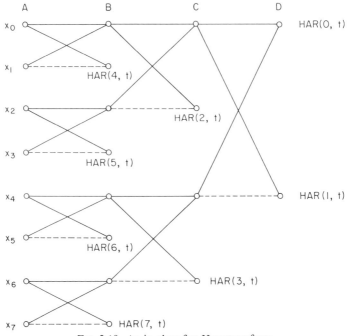

FIG. 2.19. An in-place fast Haar transform.

scribed by Roeser and Jernigan [32]. The cost of doing this is that the results will be obtained in a different order and a sorting routine will be necessary after the transformation if the rank order given in Eq. (1.71) is required. In this latter case, no improvement in speed is achieved over the previous method.

A flow diagram for this algorithm is given in Fig. 2.19, in which the exit points for individual Haar coefficients are as shown. As a consequence of the symmetric structure shown, the transformation program developed from this flow diagram can also be used for inverse transformation, provided that different initial values are used for the multiplying constants employed in the routine. In either form the Haar algorithm gives the fastest linear transformation presently available, and due to its simplified form it is particularly valuable for small computers having no floating point hardware.

Matrix relationships between the Walsh and Haar transforms have been developed by Fino [33]. He compares the subsets of Haar functions (Section 1.4) with an equivalent subdivision of the sequency-ordered Walsh functions and establishes an equivalence based on Parseval's theorem (Eq. (1.74)). From this the composite Walsh–Haar transform algorithm, shown in Fig.

2.4 The Discrete Haar Transform

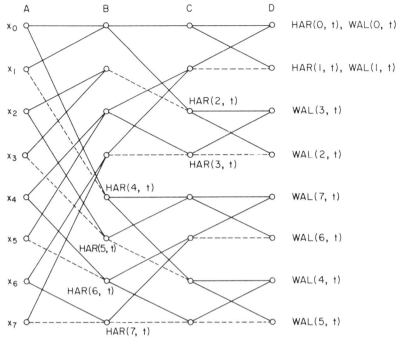

FIG. 2.20. A Walsh–Haar transform.

2.20, is derived. Here the appropriate fast-transformed coefficients are obtained at the output nodes indicated in the diagram. In the case of the Walsh output coefficients it is necessary to follow this with a reordering to obtain a sequency order. The arithmetic operations required are $2(N-1)$ for Haar (as before) and pN for Walsh (where $N = 2^p$), but this latter does not include the necessary reordering. Further generalisation is given by Fino and Algazi [34], who describe a common basic algorithm together with a set of procedural rules to allow the generation of alternative Fourier, Walsh, Haar and slant transforms. For the Haar transformation the results are obtained in natural rather than rank order (Section 1.4).

Fino's algorithm operates to produce the Haar or Walsh discrete transformations as alternatives. Hybrid transformation which combines the characteristics of the Haar and Walsh transformations has been developed by Rao et al. [35] and Huang [36]. (See also Section 1.5.) These have certain advantages in image processing and pattern recognition and are almost as efficient as the discrete Haar transform in terms of MSE and number of arithmetic operations required. The Huang algorithm, for example, requires only $2N$ additions/subtractions plus N integer multiplications.

82   2 Transformation

2.5 The discrete slant transform

The discrete slant transform was first used by Enomoto and Shibata [37] primarily as a method of bandwidth compression in television transmission systems. Although the importance of this transform has been overshadowed by the discovery of the cosine transform (see Chapter 6), it still plays an important role where a relationship is sought among different orthogonal transforms [34, 38, 39]. Jones *et al.* [39] have proved that most commonly used orthogonal transforms which exhibit half even and half odd symmetry coefficients may be obtained simply by post-processing the Walsh natural order (Hadamard) transform coefficients. Among these types of transformations the slant transform is perhaps the one most closely related to the Hadamard transform. The slant transform is another hybrid transformation in which the Walsh function series and a discrete series based on a ramp or slant waveform are combined to provide an orthogonal series whose characteristics match those found in individual scanned lines of a television image (see Subsection 1.2.6). It is due to these 'matching characteristics' that the slant transformation has the capability to compact the image energy into a comparatively small number of transformed samples.

The transform may be derived from matrix considerations, commencing with the Hadamard matrix for $N = 2$, i.e., $\mathbf{H}_2 = [\begin{smallmatrix}1 & 1\\1 & -1\end{smallmatrix}]$, which is taken as the first slant matrix $\mathbf{S}_2$ of $N \times N$ terms, with scaling factors ignored.

The slant transform matrix for $N = 4$ can be written as

$$\mathbf{S}_4 = \begin{bmatrix} 1 & 1 & 1 & 1 \\ a+b & a-b & -a+b & -a-b \\ 1 & -1 & -1 & 1 \\ a-b & -a-b & a+b & -a+b \end{bmatrix} \quad (2.40)$$

where a and b are real constants whose values are dependent on two factors:

1. S_4 must be orthogonal, and
2. the step size of the slant basis rows must be uniform throughout its length.

The first condition is achieved by making $b = 1/\sqrt{5}$, hence $a = 2/\sqrt{5}$. The second condition is achieved by making $a = 2b$. This leads to an unscaled matrix for $\mathbf{S}_4$ as

$$\mathbf{S}_4 = \begin{bmatrix} 1 & 1 & 1 & 1 \\ \dfrac{3}{\sqrt{5}} & \dfrac{1}{\sqrt{5}} & \dfrac{-1}{\sqrt{5}} & \dfrac{-3}{\sqrt{5}} \\ 1 & -1 & -1 & 1 \\ \dfrac{1}{\sqrt{5}} & \dfrac{-3}{\sqrt{5}} & \dfrac{3}{\sqrt{5}} & \dfrac{-1}{\sqrt{5}} \end{bmatrix} \quad (2.41)$$

2.5 The Discrete Slant Transform

A general expression may be developed for a slant matrix of order N in terms of a matrix of order $N/2$ in a similar manner to Hadamard matrix expansion through Kronecker multiplication, with the value of the a and b constants being determined by a recursive relationship [40]. The slant matrix so formed will possess a sequency property with a number of its middle rows being identical to a sequency-ordered Hadamard matrix. The similarity between the general form of the slant series and the sequency-ordered Walsh series can be seen when Figs. 1.4 and 1.6 are compared. A relationship between the WAL(n, t) terms (expressed with positive phasing) and the SLA(n, t) terms has been stated by Shibata [41]. For $N = 8$ it can be expressed as

$$\text{SLA}(0, t) = \text{WAL}(0, t)$$

$$\text{SLA}(1, t) = \frac{1}{\sqrt{21}}(4\ \text{WAL}(1, t) + 2\ \text{WAL}(3, t) + \text{WAL}(7, t))$$

$$\text{SLA}(2, t) = \frac{1}{\sqrt{5}}(2\ \text{WAL}(2, t) + \text{WAL}(6, t))$$

$$\text{SLA}(3, t) = \frac{1}{\sqrt{21}}\frac{1}{\sqrt{5}}(-5\ \text{WAL}(1, t) + 8\ \text{WAL}(3, t) + 4\ \text{WAL}(7, t))$$

$$\text{SLA}(4, t) = \text{WAL}(4, t)$$

$$\text{SLA}(5, t) = \text{WAL}(5, t)$$

$$\text{SLA}(6, t) = \frac{1}{\sqrt{5}}(-\text{WAL}(2, t) + 2\ \text{WAL}(6, t))$$

$$\text{SLA}(7, t) = \frac{1}{\sqrt{5}}(-\text{WAL}(3, t) + 2\ \text{WAL}(7, t))$$

(2.42)

and can be seen by combining the appropriate functions from Figs. 1.4 and 1.6 and taking note of the scaling constants and the need to invert some of the Walsh functions to achieve positive phasing.

The slant matrix is used to define the slant transform as

$$\mathbf{X}_n = \mathbf{S}_n x_i \tag{2.43}$$

where $\mathbf{S}_n$ is the generalised form of $\mathbf{S}_4$ given in Eq. (2.41) in terms of an $N \times N$ matrix.

The transform has been shown to have a good mean-square-error-coding performance for digital images, almost as good as the ideal Karhunen–Loève transform with the advantage of a fast transform algorithm which the KLT does not possess [42]. The Haar transform, discussed earlier, whilst having a more efficient transform algorithm results in a large coding error

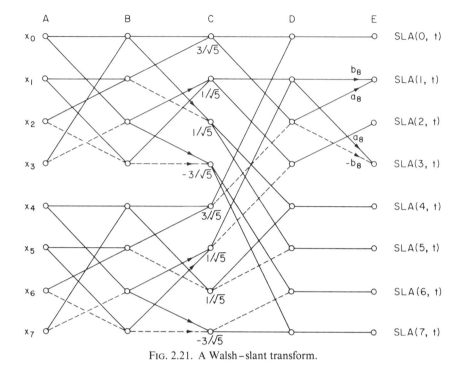

FIG. 2.21. A Walsh–slant transform.

when used for image transmission since it fails to match the characteristic features of a quantised image line.

2.5.1 Fast slant transform algorithms

The slant matrix S_n can be decomposed into a product of a series of sparse matrices. This leads to the development of a fast slant transform algorithm, viz.,

$$S_4 = \begin{bmatrix} 1 & 0 & 0 & 0 \\ 0 & \dfrac{3}{\sqrt{5}} & 0 & 0 \\ 0 & 0 & 1 & 0 \\ 0 & 0 & 0 & \dfrac{3}{\sqrt{5}} \end{bmatrix} \cdot \begin{bmatrix} 1 & 1 & 0 & 0 \\ 0 & 0 & 1 & \dfrac{1}{3} \\ 1 & -1 & 0 & 0 \\ 0 & 0 & \dfrac{1}{3} & -1 \end{bmatrix} \cdot \begin{bmatrix} 1 & 0 & 0 & 1 \\ 0 & 1 & 1 & 0 \\ 1 & 0 & 0 & -1 \\ 0 & 1 & -1 & 0 \end{bmatrix} \quad (2.44)$$

A flow diagram for $N = 8$ is shown in Fig. 2.21. Note that this algorithm requires $p + 1$ stages ($N = 2^p$) rather than p and is consequently slower than the Walsh transform. For example, where $N = 4$ the slant transform requires

2.5 The Discrete Slant Transform

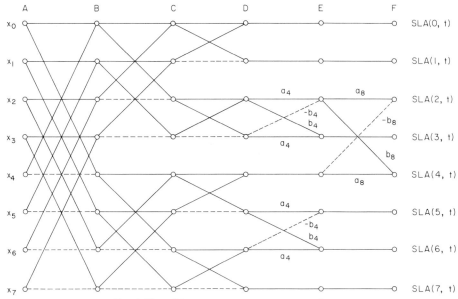

FIG. 2.22. A Wang post-processor slant transform.

8 additions/subtractions and 6 multiplications whereas the fast Walsh transform requires only 8 additions/subtractions.

A simpler algorithm requiring one more stage is due to Wang [43]. A flow diagram for this is shown in Fig. 2.22. The first three steps of this are seen to be identical to the FHT shown in Fig. 2.6 which requires $N \log_2 N$ additions/subtractions. The last two steps convert this transformation to a slant transform and require a further $(N-1)$ additions/subtractions and $2(N-2)$ multiplications. Thus in comparison with the previous method for $N=4$ this requires 10 additions/subtractions and 4 multiplications.

An FST based on the Haar series is described by Fino and Algazi [38], who combine elements of the Haar matrix with slant functions. Mathematical relationships are given which link the slant Walsh transform with the slant Haar transform. A flow diagram for this is given in Fig. 2.23. It requires $p+2$ stages, one more than the fast slant Walsh algorithm, but fewer operations are needed and it is simpler to apply since a number of self-contained butterfly routines are present. It will, however, present the slant Haar coefficients in an unordered form. An alternative algorithm is also given by Fino and Algazi [38] in which the outputs are in rank order but which will require additional storage for intermediate calculations. The required number of operations is similar in the two cases, and for $N=4$ the slant Haar transform requires 10 additions/subtractions and 2 shifts.

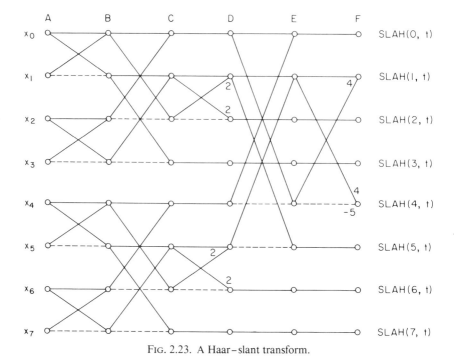

FIG. 2.23. A Haar–slant transform.

2.6 Shift-invariant transformation

A difficulty in the practical use of the Walsh transform, noted in Subsection 2.2.2, is that it is not invariant with phase shifts of the input signal as is the case with Fourier transformation. The effects of this were given in Fig. 2.1 and can be shown to inhibit the interpretation of results in such application areas as pattern recognition and image processing.

A shift-invariant transform, known as the R transform, has been described by Reitboeck and Brody [44] to overcome this limitation. Its major property is that transformed data are independent of cyclic shifts of the input signal. However, some reduction in the range of information concerning the sequency distribution of the upper range of coefficients is obtained and the transform does not provide its own inverse. Its principal value lies in pattern recognition, and since it may be developed through the Hadamard matrix a fast transform can be derived on the basis of one of the fast Walsh transforms described previously.

Invariance is obtained quite simply by replacing the subtractive terms obtained stage by stage at each node of the fast transform algorithm by their absolute values. It is necessary, however, to choose an algorithm in which the sample pairs are obtained from coefficients spaced $N/2$ values apart.

The earliest algorithm of this type was described by Ulman [23] and is given in Fig. 2.13. Here either a sequency-ordered Walsh transform or the R transform can be obtained through a simple change in one Fortran instruction to replace normal summation of a pair of coefficients by absolute summation for those intermediate computations involving a subtraction.

An in-place algorithm is described by Kunt [11] which requires fewer summations than Ulman's algorithm and therefore is faster. Here a natural-ordered Walsh transform based on the FFT algorithm (Fig. 2.7) is preceded by a bit-reversed ordering sequence and followed by a Gray code reordering. The output is in sequency order and an alternative R transformation is obtained by changing one Fortran instruction as with the Ulman program.

In both of these methods the invariant characteristic is obtained by making the functions appearing at a calculation node either $f_1(a, b) = a + b$ or $f_2(a, b) = |a - b|$. These are known as *symmetric functions*. Other symmetric functions in a, b are possible, e.g., $f(a, b) = \max(a, b)$, $\min(a, b)$, $a^2 + b^2$, $a \cdot b$, etc. Moreover, when a and b are binary values, logical functions such as $a \cdot \text{AND} \cdot b$, $a \cdot \text{OR} \cdot b$, etc., are also symmetric functions.

Wagh and Kanetkar [45] describe a general form of transform involving a range of symmetric functions of which the R transform is one. Another of these is one based on the logical functions $a \cdot \text{AND} \cdot b$ and $a \cdot \text{OR} \cdot b$, which is termed the M transform, and a fast transform algorithm for this is proposed. Since it is always much faster to carry out AND/OR operations on a digital computer compared with addition/subtraction calculations, the M transform is considerably faster to calculate than is the R transform. Other advantages in pattern recognition are claimed by Wagh and Kanetkar.

A feature of this class of fast invariant transformations is the constant geometry of the flow diagram which simplifies software and hardware implementation.

The flow diagram is shown in Fig. 2.24, in which the pairs of symmetric functions are indicated by f_1 and f_2.

Strictly speaking, the power spectrum Walsh transform which is based on $f(a, b) = a^2 + b^2$ should be included here, but for a number of reasons it is convenient to defer its consideration to the next chapter.

2.7 Transform conversion

From our brief study in this chapter of fast transform algorithms we can detect a number of similarities in the manner of their evaluation, and it is reasonable to expect that conversion from one to another should be possible through a logical algorithmic process. Indeed we have already seen how a fast Walsh transform can be obtained from a modified Cooley–Tukey FFT and how a generalised transform can be used to enable Fourier, Walsh, Haar and

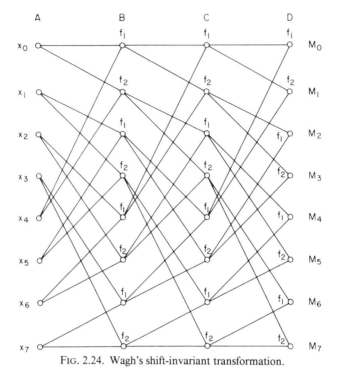

FIG. 2.24. Wagh's shift-invariant transformation.

slant transformations to be obtained by selection of suitable procedural rules or multiplying constants [15, 16, 34]. By considering the matrix partitioning characteristics of orthogonal transforms in general, Jones et al. [39] have proved that most commonly used orthogonal transforms which possess one-half of their matrix vectors in even form and the other half in odd form can be computed by sparse-matrix multiplication of any other even–odd transform. Several transforms such as the DFT, DWT, WHT, DCT, DST, SLT and HT belong to this category.

The simplest conversion is usually realised if we start with the WHT when any of the other even–odd transforms can then be achieved by carrying out a matrix post-processing of the WHT coefficients. The derivation of the SLT by Wang [43], discussed earlier, is of this type as are some of the derivations of the DCT and DST to be described in Chapter 6.

A conversion of the WHT to obtain the DFT performed by Tadokoro and Higuchi [50, 51] will be considered later in this chapter. A particular advantage of these derivations from the WHT is that since this algorithm is easy to implement either by software or in hardware, and provided that the post-processing operations are not too complex, an overall saving in computer time or hardware can be realised.

2.7 Transform Conversion

Other methods of obtaining a transform conversion have been carried out, and a number of methods have been developed for the conversion of a set of DWT coefficients to obtain an equivalent DFT. These rely on the relationship between the Fourier and Walsh function series given in Subsection 1.3.6 and have been applied by Kitai [46] and others [45, 47] to obtain a fast transform procedure.

In one of these, due to Abramson [47], a staircase approximation of the function is converted to a set of sinc and cosinc terms in the frequency domain that is equivalent to the Fourier transform. The sinc and cosinc terms are given as

$$\text{sinc } x = \frac{\sin \pi x}{\pi x}, \quad \text{cosinc } x = \frac{1 - \cos \pi x}{\pi x} \quad (2.45)$$

which have been shown to be easier to express in discrete Walsh terms than the sine–cosine terms.

The process first approximates the waveform in terms of Walsh functions and then proceeds to determine the equivalent Fourier transformation through the sinc and cosinc transforms expressed in sequency terms, viz.,

$$ST(f) = \sum_{k=1}^{\infty} X_c(k)a(ck), \quad CT(f) = \sum_{k=1}^{\infty} X_s(k)b(sk) \quad (2.46)$$

where $X_c(k)$ and $X_s(k)$ are the CAL- and SAL-transformed values and $a(ck)$ and $b(sk)$ the cosine and sine transforms of the CAL and SAL functions, respectively.

Since the fast Fourier transform is not involved, there are no aliasing problems. This can be advantageous in certain circumstances where with the FFT an error is produced by sampling at less than the Nyquist rate. Normally to reduce this the sampling interval would have to be reduced. However, by using the sinc transform with its square-wave kernel, this error does not occur. In effect the interpolation and scaling in the frequency domain can be regarded as 'built in', so that any frequency point may be chosen without changing the number or spacing of the samples in the time domain.

Instead of calculating $X_c(k)$ and $X_s(k)$ separately and finding the summation of the products with $a(ck)$ and $b(sk)$, it has been shown to be considerably quicker to determine the intervening set of sinc–cosinc coefficients through the calculation of product of a sparse matrix **S**, which combines the transposition of a Walsh matrix and a sinc–cosinc array [48]. A fast transform algorithm to do this requires $2(N - 1)$ addition or subtraction operations compared with $N \log_2 N$ for the fast Walsh transform.

The discrete Fourier transform (neglecting the zero frequency term) is finally obtained from the summation

$$ST(f) + CT(f) = \frac{1}{2\pi f} x_i \cdot \mathbf{S} \cdot \mathbf{V} \quad (2.47)$$

where **V** is a previously calculated sine–cosine vector containing a set of frequency-related terms.

Calculation of Eq. (2.47) is slower than the FFT but is devoid of aliasing. Abramson [47] has suggested that the evaluation of a set of **S** coefficients, which is faster than the FWT, could be used alone for image data compression in place of either Fourier or Walsh coefficients.

Two further methods of Fourier-from-Walsh transformation have been developed. One of these, due to Siemans and Kitai [49], is suitable for hardware instrumentation and has been developed as a flexible method of spectral analysis. The second, which is described in two slightly different versions, is due to Tadokoro and Higuchi [50–52] and ofers a fast software method of Fourier coefficient determination which can be considerably faster than the FFT.

In Kitai's method the signal of length T is first transformed (by hardware or computer software) into a set of CAL and SAL coefficients. These are then subject to a multiplicative and summation process to arrive at the Fourier cosine and sine coefficients, viz.,

$$a_n = \frac{2}{T} \int_0^T \text{CAL}(k, t') \cos n\omega_0 t' \, dt'$$
$$b_n = \frac{2}{T} \int_0^T \text{SAL}(k, t') \sin n\omega_0 t' \, dt'$$
(2.48)

where $k = 1, 2, 3, \ldots$ and $t' =$ normalised time $= t/T$.

In this method the Walsh function coefficients are obtained through a synchronised hardware Walsh spectral analyser and the corresponding Fourier series obtained through a digital computer calculation carried out after the measurement has been completed. The values for the sine and cosine terms are computed previously and stored in memory as a *look-up table*.

Advantages of obtaining a Fourier analysis in this way derive from the hardware simplicity and speed of Walsh spectral analysers. The multiplication by sines and cosines in Fourier processing is obviated in Walsh processing, where signal values may be simply accumulated with or without sign change, depending on the signs of the signal and its SAL or CAL values at any instant. By using this method it has also been proved possible to extend the frequency measurement to low frequencies of the order of several days per hertz without excessive averaging time.

A fundamental difficulty in using this technique, however, lies in the fact that frequency-limited signals have an infinite Walsh spectra and the set of Walsh coefficients derived will necessarily be a truncated one. This can introduce considerable error in the Walsh-to-Fourier conversion. In a later development Siemans and Kitai [53] introduced a correction factor to minimise this souce of error. Here the compensation for truncation consists of a diagonal matrix that pre-multiplies the Walsh-to-Fourier conversion, which can also be considered in matrix terms. Thus for frequency-limited periodic signals Siemans and Kitai show that the error-corrected Fourier coefficients **a** and **b** replacing a_n and b_n of Eq. (2.48) can be expressed as

$$\mathbf{a} = \mathbf{D}_a \Lambda_a \mathbf{A} \quad \text{and} \quad \mathbf{b} = \mathbf{D}_b \Lambda_b \mathbf{B} \tag{2.49}$$

where **A** and **B** are vectors whose elements are the derived CAL and SAL coefficients, $\mathbf{D}_a$ and $\mathbf{D}_b$ diagonal correction matrices and Λ_a and Λ_b square matrices whose elements are the Fourier coefficients of Walsh functions.

The compensation matrices consists of elements of sinc = $\sin(x)/x$ terms derived for each Fourier coefficient of normalised frequency, viz.,

$$\begin{aligned} \mathbf{D}_{af} = \mathbf{D}_{bf} &= \operatorname{sinc}^{-2}(f/2^n) \quad \text{for} \quad f < 2^{n-1} \\ \mathbf{D}_{bf} &= 1 \cdot 23 \quad \text{for} \quad f = 2^{n-1} \end{aligned} \tag{2.50}$$

Various methods of expression for the Fourier coefficients of a Walsh series are available. In this derivation the non-recursive Gray code representation is used [53]. For a given sequency range a set of Λ_a and Λ_b values can be pre-calculated and used in subsequent calculations, as can the set of K values for a given matrix size (value of N). Thus the only measured coefficients obtained directly from the signal are the CAL and SAL values from the truncated Walsh transformation.

Tadokoro and Higuchi's method is less complex computationally and can be considerably faster to calculate than the FFT, particularly if all possible coefficients are not needed. In their original paper [50] a necessary condition for analysis was that the signal $x(t)$ be sampled at points shifted by π/N degrees from the conventional sampling points in the DFT, considering the orthogonality between Walsh and sinusoidal functions (see Figs. 1.1 and 1.4). In the modified procedure described by Tadokoro and Higuchi [51] this restruction is removed and it is this latter method that is summarised here.

As with the earlier methods the similarities between the sine–cosine and SAL–CAL functions are noted and a relationship between the two series representing the same sampled signal is found. In Fourier terms the signal x_i can be represented as

$$x_i = a_0 + \sum_{k=1}^{N/2-1} (a_k \cos(k\theta_i) + b_k \sin k\theta_i) + a_{N/2} \cos((N/2)\theta_i) \tag{2.51}$$

where $\theta_i = 2\pi i/N$ and $i = 0, 1, \ldots, N-1$. A column vector $\mathbf{F}_N$ is obtained from these sampled values. We can obtain a Walsh matrix $\mathbf{W}_N$ by appropriately sampling the Walsh function series WAL(n, t) for $N = 2^p$ and relate $\mathbf{F}_N$

and $\mathbf{W}_N$ through the DWT, viz.,

$$\mathbf{A}_N = \frac{1}{N} \cdot \mathbf{W}_N \cdot \mathbf{F}_N \qquad (2.52)$$

If the right-hand side of Eq. (2.52) is evaluated in terms of its Fourier coefficients from Eq. (2.51), then a set of simultaneous equations of the first order can be derived for a_k and b_k.

This is the point at which the method differs markedly from that of Siemans and Kitai. Whereas in their method matrix calculation is used to extract a_k and b_k, in Tadokoro and Higuchi's method the values of a_k and b_k are obtained through the solution of a set of simultaneous equations in A_N.

A simple example for $N = 8$ illustrates this latter approach. The eight A_N Walsh coefficients are evaluated by using the FWT. This gives the left-hand side of Eq. (2.52). The right-hand side is evaluated for eight sampled terms to give

$$\begin{bmatrix} A_0 \\ A_1 \\ A_2 \\ A_3 \\ A_4 \\ A_5 \\ A_6 \\ A_7 \end{bmatrix} = \begin{bmatrix} a_0 \\ (a_1 + a_3 + (1 + \sqrt{2})b_1 + (-1 + \sqrt{2})b_3)/4 \\ ((1 + \sqrt{2})a_1 + (1 - \sqrt{2})a_3 - b_1 + b_3)/4 \\ (a_2 + b_2)/2 \\ (a_2 - b_2)/2 \\ ((1 - \sqrt{2})a_1 + (1 + \sqrt{2})a_3 - b_1 + b_3)/4 \\ (a_1 + a_3 + (1 - \sqrt{2})b_1 + (-1 - \sqrt{2})b_3)/4 \\ a_4 \end{bmatrix} \qquad (2.53)$$

Solving for a_k and b_k gives

$$\begin{bmatrix} a_0 \\ a_1 \\ b_1 \\ a_2 \\ b_2 \\ a_3 \\ b_3 \\ a_4 \end{bmatrix} = \begin{bmatrix} A_0 \\ ((A_1 + A_2 + A_5 + A_6)/2 + (A_2 - A_5))/\sqrt{2} \\ ((A_1 - A_2 - A_5 + A_6)/2 + (A_1 - A_6))/\sqrt{2} \\ A_3 + A_4 \\ A_3 - A_4 \\ ((A_1 + A_2 + A_5 + A_6)/2 - (A_2 - A_5))/\sqrt{2} \\ -((A_1 - A_2 - A_5 + A_6)/2 - (A_1 - A_6))/\sqrt{2} \\ A_7 \end{bmatrix} \qquad (2.54)$$

Two features are apparent from Eq. (2.54). First each Fourier coefficient is represented from linear combinations of values of additions and subtractions among the Walsh coefficients. Second, the combinations repeat with inversion of sign for certain pairs of Fourier coefficients (e.g., $a_1, a_3; b_1, b_3; a_2, b_2$). These facts make possible an economical algorithm for obtaining the set of these linear combinations. This is shown for $N = 8$ in the flow diagram of Fig. 2.25. The alternative inputs at stage 2 are shown for clarity by switches S_1 and S_2, but in practice software selection would be used. From this

2.8 Two-Dimensional Transformation

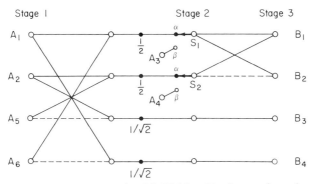

FIG. 2.25. Tadokoro and Higuchi's Walsh-to-Fourier transformation.

diagram we see that

$$a_1 = B_{1\alpha} + B_3, \qquad b_1 = B_{2\alpha} + B_4$$
$$a_2 = B_{1\beta}, \qquad b_2 = B_{2\beta} \qquad (2.55)$$
$$a_3 = B_{1\alpha} - B_3, \qquad b_3 = -B_{1\alpha} - B_4$$

where $B_{1\alpha}$ indicates B_1 output with switch S_1 in the α position and $B_{1\beta}$ for the β position, with $B_{2\alpha}$ and $B_{2\beta}$ designated similarly. Also, a_0 and a_4 are obtained directly from A_0 and A_7, respectively. The complete algorithm is described by Tadokoro and Higuchi [51], who show that the total number of multiplications required is approximately $NL/9$, where N is the number of data values and L the number of Fourier coefficients desired. This is fewer than the multiplications required by the FFT for $N < 256$ when all the coefficients are obtained. Where relatively few coefficients are required (e.g., $L \ll N$) as found in certain forms of spectral analysis and other applications, a further improvement is obtained. A full description of the calculation equations and their derivation is given by Tadokoro and Higuchi [52].

2.8 Two-dimensional transformation

For image processing the sampled image is available as an $N \times N$ array of discrete values, and domain transformation is generally carried out in terms of a single-dimensional transform algorithm applied sequentially to the rows and columns of the matrix array.

The two-dimensional finite Walsh transform of a two-dimensional array $x_{i,j}$ of N^2 values is given by

$$X_{m,n} = \frac{1}{N^2} \sum_{i=0}^{N-1} \sum_{j=0}^{N-1} x_{i,j} \, \text{WAL}(m, j) \qquad (2.56)$$

and the inverse transformation by

$$x_{i,j} = \sum_{m=0}^{N-1} \sum_{n=0}^{N-1} X_{m,n} \, \text{WAL}(n, i) \, \text{WAL}(m, j) \qquad (2.57)$$

using the symmetry of the discrete Walsh transform.

Transformation may be carried out in two steps. First the transformation for the variable i is performed, viz.,

$$X_{m,j} = \sum_{i=0}^{N-1} x_{i,j} \, \text{WAL}(n, i) \qquad (2.58)$$

This is equivalent to a one-dimensional transformation along each row of the array. Then a second one-dimensional transform is taken along each column of the transformed array for the variable j, viz.,

$$X_{m,n} = \sum_{j=0}^{N-1} X_{m,j} \, \text{WAL}(m, j) \qquad (2.59)$$

The process is illustrated for the FWT on an 8 × 8 image in Fig. 2.26. First a fast Walsh transformation is taken of each row of the image data and then followed by a fast transformation of each column of the obtained matrix. The number of operations required is $2N^2 \log_2 N$ compared with $2N^3$ for the direct application of Eq. (2.56). An alternative method due to Bates [54] which also uses a single-dimensional transformation to derive a two-dimensional transform is to rearrange the image matrix into a one-dimensional

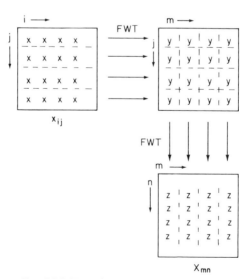

FIG. 2.26. Two-dimensional transformation.

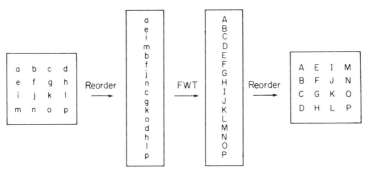

FIG. 2.27. Alternative method of two-dimensional transformation.

vector with the columns following on sequentially (see Fig. 2.27). A one-dimensional transform of N^2 values is taken of the vector and the resulting output rearranged again in matrix form.

As with the single-dimensional transform Parseval's relationship can be applied, giving

$$\sum_{i=1}^{N-1}\sum_{j=0}^{N-1}|x_{i,j}|^2 = \frac{1}{N^2}\sum_{m=0}^{N-1}\sum_{n=0}^{N-1}|X_{m,n}|^2 \qquad (2.60)$$

This has important implications in bandwidth reduction since if a few of the Walsh domain samples are of a large magnitude, then it follows that the remainder will be of a small magnitude. These small-magnitude samples may be discarded to achieve a reduced bandwidth for the data sample. This is the method of threshold filtering discussed in Chapter 3.

A further discussion on two-dimensional transformation is given in Chapter 6, where the various orthogonal transforms used for this purpose will be reviewed.

References

1. Good, I. J. (1958). The interactive algorithm and practical Fourier analysis. *J. Roy. Stat. Soc. (London)* **B20**, 361–372.
2. Cooley, J. W., and Tukey, J. W. (1965). An algorithm for the machine computation of complex Fourier series. *Math. Comput.* **19**, 297–301.
3. Gentleman, W. M., and Sande, G. (1966). Fast Fourier transforms for fun and profit. *AFIPS Proc. Fall Joint Comp. Conf.* **29**, 563–578.
4. Welchel, J. E., and Guinn, D. F. (1968). The fast Fourier–Hadamard transform and its use in signal representation and classification. Tech. Report PRC68-11, Melpar, Inc., Falls Church, Virginia.
5. Pichler, F. (1970). Some aspects of a theory of correlation with respect to Walsh harmonic analysis. Univ. of Maryland Report R-70-11, College Park, Maryland.
6. Pichler, F. (1967). Das System der sal und cal Funktionen als Erweiterung des Systems der

Walsh Funktionen und die Theorie der sal und cal Fourier Transformation. Ph.D. thesis, Univ. of Innsbruck, Innsbruck, Austria.
7. Swick, D. A. (1969). Walsh function generation. *IEEE Trans. Comput.* **C-20,** 211–213.
8. Pratt, W. K., Kane, J., and Andrews, H. C. (1969). Transform image coding. *Proc. IEEE* **57,** 58–68.
9. Yuen, C. K. (1972). Remarks on the ordering of Walsh functions. *IEEE Trans. Comput.* **C-21,** 1452.
10. Shanks, J. L. (1969). Computation of the fast Walsh–Fourier transform. *IEEE Trans. Comput.* **C-18,** 457–459.
11. Kunt, M. (1975). On computation of the Hadamard transform and the R transform in ordered form. *IEEE Trans. Comput.* **C-24,** 1120–1121.
12. Manz, J. W. (1972). A sequency-ordered fast Walsh transform. *IEEE Trans. Audio Electroacoust.* **AV-20,** 204–205.
13. Larsen, H. (1976). An algorithm to compute the sequency-ordered Walsh transform. *IEEE Trans. Acoust., Sp. Sig. Proc.* **ASSP-24,** 335–336.
14. Caspari, K. (1970). Generalised spectrum analysis. *Proc. Applic. Walsh Functions, Washington, D.C.,* AD707431, pp. 195–207.
15. Ahmed, N., and Rao, K. R. (1971). The generalised transform. *Proc. Applic. Walsh Functions, Washington, D.C.,* AD727000, pp. 60–67.
16. Madych, W. R. (1978). Generalised Walsh-like expansions. *IEEE Midwest Symp. Circ. Syst., 21st,* pp. 378–392.
17. Kunt, M., and de Coulan, F. (1973). On generalised fast transformations in digital signal processing. *NTG Conf. Inf. and Syst. Theory, Erlangen, Germany,* pp. 294–311.
18. Vlazenko, V., and Rao, K. R. (1978). A generalised approach to orthogonal transform algorithms. *IEEE Midwest Symp. Circ. Syst., 21st,* pp. 388–392.
19. Fontaine, A. B. (1973). Simple dyadic and sequency Fortran mechanisations of the fast Walsh transform. *Proc. Nat. Electron. Conf.* **28,** 271–273.
20. Blackman, R. B., and Tukey, J. W. (1958). "Measurement of Power Spectra." Dover, New York.
21. Brown, R. D. (1977). A recursive algorithm for sequency-ordered fast Walsh transforms. *IEEE Trans. Comput.* **C-26,** 819–822.
22. Harmuth, H. F. (1972). "Transmission of Information by Orthogonal Functions." Springer-Verlag, Berlin.
23. Ulman, L. J. (1970). Computation of the Hadamard transform and the R-transform in ordered form. *IEEE Trans. Comput.* **C19,** 359–360.
24. Berauer, G. (1972). Fast 'in-place' computation of the discrete Walsh transform in sequency order. *Proc. Symp. Applic. Walsh Functions, Washington, D.C.,* AD744650, pp. 272–274.
25. Muniappan, K., and Kitai, R. (1982). Walsh spectrum measurement in natural, dyadic and sequency ordering. *IEEE Trans. Electromag. Compat.* **EMC-24,** 46–49.
26. Carl, J. W., and Swartwood, R. V. (1973). A hybrid Walsh transform computer. *IEEE Trans. Comput.* **C-22,** 669–672.
27. Smith, E. G. (1980). Computer-aided design of a PLA implemented fast Walsh–Hadamard transform device. *IEEE Proc. Int. Conf. Pattern Recognition, 5th,* pp. 183–191.
28. Geadah, Y. A., and Corinthios, M. J. G. (1977). Natural, dyadic and sequency order algorithms and processors for the Walsh–Hadamard transform. *IEEE Trans. Comput.* **C-26,** 435–442.
29. Rao, K. R., Devarajan, V., Vlasenko, V., and Narasimhan, M. A. (1978). CAL–SAL Walsh–Hadamard transform. *IEEE Trans. Acoust., Sp. Sig. Proc.* **ASSP-26,** 605–607.
30. Kunt, M. (1979). In-place computation of the Hadamard transform in CAL–SAL order. *Signal Processing,* 1, 227–231.

31. Rejchrt, V. J. (1972). Signal flow graph and a Fortran program for the Haar–Fourier transform. *IEEE Trans. Comput.* **C-21,** 1026–1027.
32. Roeser, P. R., and Jernigan, M. E. (1982). Fast Haar transform algorithms. *IEEE Trans. Comput.* **C-31,** 175–177.
33. Fino, B. J. (1972). Relations between Haar and Walsh–Hadamard transforms. *Proc. IEEE (Lett.)* **60,** 647–648.
34. Fino, B. J., and Algazi, V. R. (1977). A unified treatment of discrete fast transforms. *SIAM J. Computing* **6,** 700–717.
35. Rao, K. R., Narasimhan, M. A., and Revuluri, K. (1975). Image data processing by Hadamard–Haar transform. *IEEE Trans. Comput.* **C-24,** 888–896.
36. Huang, D. M. (1980). Walsh–Hadamard–Haar hybrid transforms. *IEEE Proc. Int. Conf. Pattern Recognition, 5th,* pp. 180–182.
37. Enomoto, H., and Shibata, K. (1971). Orthogonal transforms coding system for television signals. *Proc. Symp. Applic. Walsh Functions, Washington, D.C.,* AD727000, pp. 11–17.
38. Fino, B. J., and Algazi, V. R. (1974). Slant–Haar transform. *Proc. IEEE* **62,** 653–654.
39. Jones, H. W., Hein, D. N., and Knauer, S. C. (1978). The Karhunen–Loève discrete cosine and related transforms obtained via the Hadamard transform. *Int. Telemetering Conf. U.S.A., Los Angeles,* pp. 87–98.
40. Pratt, W. K., Welch, L. R., and Chen, W. H. (1972). Slant transforms for image coding. *Proc. Symp. Applic. Walsh Functions, Washington, D.C.,* AD744650, pp. 229–234.
41. Shibata, K. (1972). Waveform analysis of image signals by orthogonal transformation. *Proc. Symp. Applic. Walsh Functions, Washington, D.C.,* AD744650, pp. 210–215.
42. Pratt, W. K., Chen, W. H., and Welch, L. R. (1974). Slant transform image coding. *IEEE Trans. Commun.* **COM-22,** 1075–1093.
43. Wang, Z. D. (1982). New algorithm for the slant transform. *IEEE Trans. Pattern Anal. Mach. Intell.* **PAMI-4,** 551–555.
44. Reitboeck, H., and Brody, T. P. (1968). A transformation with invariance under cyclic permutation for application in pattern recognition. *Info. Cont.* **15,** 130–154.
45. Wagh, M. D., and Kanetkar, S. V. (1977). A class of translation invariant waveforms. *IEEE Trans. Acoust. Sp. Sig. Proc.* **ASSP-25,** 203–205.
46. Kitai, R. (1975). Walsh to Fourier spectral conversion for periodic waves. *IEEE Trans. Electromag. Compat.* EMC-17, 266–269.
47. Abramson, R. F. (1977). The sinc and cosinc transform. *IEEE Trans. Electromag. Compat.* **EMC-19,** 88–94.
48. Bates, R. H. T., Napier, P. J., and Chang, Y. P. (1970). Square-wave Fourier transforms. *Electron. Lett.* **6,** 741–742.
49. Siemans, K. H., and Kitai, R. (1969). Digital Walsh–Fourier analysis of periodic waveforms. *IEEE Trans. Instrum. Meas.* **IM-18,** 316–321.
50. Tadokoro, Y., and Higuchi, T. (1978). Discrete Fourier transform computation via the Walsh transform. *IEEE Trans. Acoust., Sp. Sig. Proc.* **ASSP-26,** 236–240.
51. Tadokoro, Y., and Higuchi, T. (1981). Another discrete Fourier transform with small multiplications via the Walsh transform. *IEEE Int. Conf. Acoust., Sp. Sig. Proc.* **81,** 306–309.
52. Tadokoro, Y., and Higuchi, T. (1983). Conversion factors from Walsh coefficients to Fourier coefficients. *IEEE Trans. Acoust., Sp. Sig. Proc.* **ASSP-31,** 231–232.
53. Siemans, K. H., and Kitai, R. (1973). A non-recursive equation for the Fourier transform of a Walsh function. *IEEE Trans. Electromag. Compat.* **EMC-15,** 81–83.
54. Bates, R. M. (1971). Multi dimensional BIFORE transform. Ph.D. dissertation, Kansas State University, Manhattan, Kansas.

Chapter 3

Analysis and Processing

3.1 Introduction

The transformation and generation processes described in the preceding chapters form the basis of a number of digital analysis and processing procedures which find use in practical applications of sequency functions. These procedures consist principally of spectral decomposition and synthesis and correlation and digital filtering for both single- and multi-dimensional signals. The various methods for achieving these operations will be discussed in this chapter and comparisons made with existing Fourier techniques.

Whilst similarities may be seen with appropriate Fourier operations, there are also considerable differences, which lead in some cases to processing possibilities which do not have a Fourier counterpart. The sequency function should therefore be considered as a processing tool in its own right and not simply as a different or more convenient way of carrying out routine analysis and processing operations. If this is not done, the opportunities of matching fully these newer sequency functions to the application being investigated may be missed.

One area in which these differences are most pronounced lies in the treatment of correlation between signals and in autocorrelation. This is probably the least successful area for the application of sequency functions since a major problem arises with unsynchronised signals due to the absence of a shift theorem for the functions which we find with the sine and cosine functions.

This shortcoming pertains particularly in the derivation of the power spectral density through the transformed product of correlation functions, as we shall see later. Other forms of power spectral evaluation using sequency functions can, however, be obtained more rapidly than the corresponding Fourier analysis, with consequent savings in computational cost. The use of sequency functions in various matching and digital filtering operations can also provide useful processing characteristics for the least complexity. This applies particularly in image processing, where the numbers of samples to be processed is large. Finally, the use of sequency functions in waveform synthesis has led to a number of generation procedures in which the simplicity of the functions and their convergence properties favours their use over other methods.

3.2 Correlation and convolution

In correlation and convolution the behaviour of the Walsh and other sequency functions is so different from the sine–cosine functions that a different view of a series of time lag values to that experienced in Fourier analysis is required. We shall commence with a consideration of the convolution of two Walsh series which illustrates the special properties of convolution and correlation in the sequency domain.

Let us define two time series x_i and y_i, which have transformed values in the Walsh domain as X_k and Y_k, respectively. The convolution of the two time series can be expressed as

$$Z(\tau) = \frac{1}{N} \sum_{i=0}^{N-1} x_i y_{\tau-i} = X_k * Y_k \qquad (3.1)$$

where τ indicates the incremental lag value.

Replacing x_i and $y_{\tau-i}$ with their transformed values and letting $Y_k = Y_l$

$$Z(\tau) = \frac{1}{N} \sum_{i=0}^{N-1} \left[\sum_{k=0}^{N-1} X_k \, \text{WAL}(k, i) \right] \left[\sum_{l=0}^{N-1} Y_l \, \text{WAL}(l, \tau - i) \right]$$

$$= \sum_{k=0}^{N-1} \sum_{l=0}^{N-1} X_k Y_l \left[\frac{1}{N} \sum_{i=0}^{N-1} \text{WAL}(k, i) \, \text{WAL}(l, \tau - i) \right] \qquad (3.2)$$

The summation of Walsh term products shown within the brackets indicate the convolution of the discrete Walsh functions.

If we carry out a similar convolution by using the Fourier transform, the corresponding time delay function $Y(\tau - t) \exp(-j2\pi i n/N)$ can be decomposed into sine–cosine products by the use of the *shift theorem* to permit the simple relationship

$$x(t) * y(t) \leftrightarrow X(f)Y(f) \qquad (3.3)$$

to be applied, where ↔ indicates a transform operator. Thus, convolution of $x(t)$ and $y(t)$ becomes equivalent to the inverse transform of the product of their Fourier transforms $X(f)$ and $Y(f)$.

No such simple relationship between a Walsh function and a delayed version of the same function exists, and so a direct equivalence between the products of the Walsh transforms and the convolution of the time domain representations cannot be used. The same consideration also prevents any simple expression for auto- or cross correlation in terms of the Walsh transform coefficients.

However, by making use of the Walsh *addition theorem* (Eq. (2.15)), we find that this can play a similar part to the shift theorem in Fourier convolution. Thus, for two series

$$x_i \leftrightarrow X_k, \qquad y_i \leftrightarrow Y_k \tag{3.4}$$

we are able to define *dyadic convolution* (also known as *logical convolution* [1]) as

$$Z(\tau) = \frac{1}{N} \sum_{i=0}^{N-1} x_i y_{\tau \oplus i} = X_k \circledast Y_k \tag{3.5}$$

where ⊛ represents the operation of dyadic convolution and

$$Z(\tau) = \frac{1}{N} \sum_{i=0}^{N-1} x_i \left[\sum_{k=0}^{N-1} Y_k \, \text{WAL}(k, \tau \oplus i) \right] \tag{3.6}$$

Using Eq. (2.15),

$$Z(\tau) = \frac{1}{N} \sum_{k=0}^{N-1} Y_k \sum_{i=0}^{N-1} x_i \, \text{WAL}(k, i) \, \text{WAL}(k, \tau)$$

$$= \sum_{k=0}^{N-1} X_k Y_k \, \text{WAL}(k, \tau) \tag{3.7}$$

This results in an equivalent relationship to Eq. (3.3), viz.,

$$x_i \circledast y_i \leftrightarrow X_k Y_k \tag{3.8}$$

Hence two different sets of relationships exist side by side for Fourier and Walsh series. Both express a form of convolution theory, but, whereas the Fourier version implies arithmetic addition for the recursive time shift, the Walsh version requires the substitution of dyadic or modulo-2 addition. Using these expressions we are now in a position to compare discrete correlation in the Fourier and Walsh cases.

Discrete autocorrelation in real time may be defined as

$$R_F(\tau) = \frac{1}{N} \sum_{i=0}^{N-1} x_i x_{i+\tau} \tag{3.9}$$

3.2 Correlation and Convolution

where $i = 0, 1, 2, \ldots, m$ and $m \ll N$. Here m is the total correlation lag. Discrete autocorrelation in dyadic time is given as

$$R_W(\tau) = \frac{1}{N} \sum_{i=0}^{N-1} x_i x_{i \oplus \tau} \qquad (3.10)$$

The difference between the two expressions lies in the addition behaviour of the incremental time lag. In the real time domain this addition is arithmetic and in the Walsh domain this is modulo-2 addition. Similar remarks apply to cross correlation of two discrete time series.

Dyadic convolution and correlation are identical because modulo-2 addition and subtraction are identical operations. How then are we to correlate the Walsh series either with itself or with another Walsh series?

Rapid evaluation through the product of the Walsh transformations is not feasible for reasons given earlier. It is possible, of course, to evaluate Eq. (3.2) directly and determine the functions from the summation of sequency-ordered Walsh transforms. A computer algorithm to do this is given by Kennett [2] but is limited in usefulness since even more operations are needed than in the direct evaluation of Eq. (3.1).

There are, however, a number of other methods that may be used which, although containing certain limitations, may be used with advantage in many cases. One of these is the recursive procedure, due to Yuen [3], which is applicable for a special set of time shifts for which the delay values are reciprocal powers of two. This procedure uses a recursive equation which relates correlations of high index functions to those of low index functions. Successive applications of the equation yield expressions for evaluating the correlations for the restricted values of time shifts. A computing procedure described by Yuen [4] is approximately as fast as the evaluation of a fast Walsh transform for the same number of values. This can be used in certain applications in communications, such as orthogonal decoding and sequency demultiplexing [5, 6], but not to those problems requiring an arbitrary time shift.

Another algorithm, due to Pitassi [7], does not contain this limitation but is effectively confined to short-term series. This algorithm can actually be faster than evaluation through the product of fast Fourier transforms for the input series where $N < 1024$, but it requires a fairly large memory storage of intermediate results of $2 \times 3^{p-1}$, where $p = \log_2 N$.

Certain applications allow the use of dyadic correlation as defined in Eq. (3.5) in which the time delay values are obtained by modulo-2 addition, so that the product of two Walsh transforms may be directly employed. Examples of this are in the reception of sequency-modulated transmissions [5] and in certain forms of transform spectroscopy studies [8]. A hardware correlator based on Eq. (3.5) has been described by Frank [9]. Maqusi [10] applies dyadic correlation to systems analysis and develops a matrix method which allows factorisation for fast calculations similar to the transformation techniques discussed in Chapter 2.

3.2.1 The dyadic time scale

It was shown in the preceding section how the process of correlation for Walsh series is dependent on the time addition being carried out by using modulo-2 addition rather than arithmetic or linear lag addition. This process imparts a peculiar behaviour to time in what has become known as the *dyadic domain.* Dyadic time is compared with real (linear) time in Fig. 3.1. Here consecutive data samples for a process are shown equally spaced (that is in real time) so that, whereas in arithmetic correlation the lag increases uniformly with time, in the dyadic (Walsh) case the time lag varies in the manner shown. Consecutive data samples are taken at time instants which progress in a series of jumps unequal in length over both forward and backward time intervals.

It is difficult to associate the dyadic time scale with our normal experience of correlation, and this fact is what makes correlation results obtained by using the Walsh series of so little direct value. The results of this type of correlation obtained in dyadic time are of use, however, since conversion to arithmetic correlation can be made through a matrix translation. The procedure has certain advantages in spectral evaluation, which will be discussed in the next section.

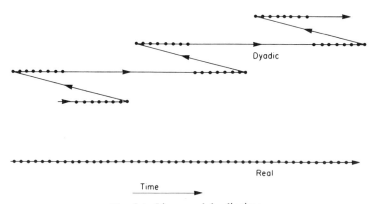

FIG. 3.1. Linear and dyadic time.

3.3 Spectral analysis

Spectral analysis in terms of sequency rather than frequency was first suggested in a paper by Polyak and Schreider [11] in which the Walsh function is applied to the spectral analysis of discontinuous and transient waveforms. A form of spectral analysis based on the Walsh series was developed by Gibbs and Millard [12], who produced a theory for Walsh functions analogous to that of Wiener–Khintchine and which leads to a definition of the power spectrum through the classical route. A *periodogram* estimation for the Walsh power spectrum has been described [13] which is closely similar to the technique developed for sinusoidal functions. These are all averaging techniques which describe the average power distribution of a random process.

Other methods take advantage of the special properties of natural Walsh and Haar ordering to define forms of power spectrum, such as those proposed by Ohnsorg [14] and developed by Ahmed *et al.* [15], to give a highly compressed spectral representation, which will be referred to here as the *harmonic spectrum*.

An important feature of the definition of the power spectrum using Walsh functions is that it is possible for the power spectrum to be sequency limited although the corresponding time functions are time limited. This feature contrasts the behaviour of the Fourier transform in a power spectrum definition in which a time-limited function cannot have a frequency-limited power spectrum. This has relevance to the application of Walsh functions in the analysis of non-stationary data as noted by Gibbs and Millard [12].

The sequency spectral decomposition of a smoothly-varying waveform, such as a sinusoid, was shown earlier to have a more complex spectrum than is obtained with Fourier analysis. This is illustrated in Fig. 3.2, which shows the pattern of sequency power distribution for a wide band of sinusoidal waveforms. For those frequencies which are not binary multiples of the sequency time base, considerable energy is found outside the sequency which has zero-crossing equality with that of the sinusoidal waveform. A similar complexity is obtained by the Fourier analysis of a rectangular waveform. Figure 3.3 is a comparison between the power spectra obtained in the two cases.

A second feature of Walsh spectral decomposition is the effect of circular time shift noted earlier (Subsection 2.2.2). As seen in Fig. 2.1, corresponding SAL and CAL functions of the same sequency will vary quite considerably in amplitude, but in a reciprocal manner. This suggests that the summation of squares of the CAL and SAL function coefficients will mitigate this effect, and indeed this procedure forms the basis of the periodogram method of

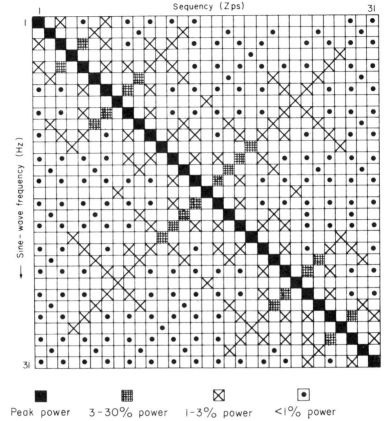

Fig. 3.2. Sequency power distribution for a set of sinusoidal waveforms of frequency 1–31 Hz over a Walsh sequency spectrum extending from 1 to 31 Zps.

power spectrum estimation. The problem of finding an acceptable level of invariance is central to any discussion of the sequency power spectrum, and a number of techniques for achieving invariance with circular time shift will be discussed in the following pages.

As with Fourier analysis three alternative classes of operation can be carried out to derive the averaging Walsh power spectrum.

(1) An indirect method via the dyadic autocorrelation function (equivalent to the Wiener–Khintchine method,

(2) direct evaluation via the squared value of the Walsh transform (equivalent to the periodogram method), and

(3) narrow-band Walsh filtering.

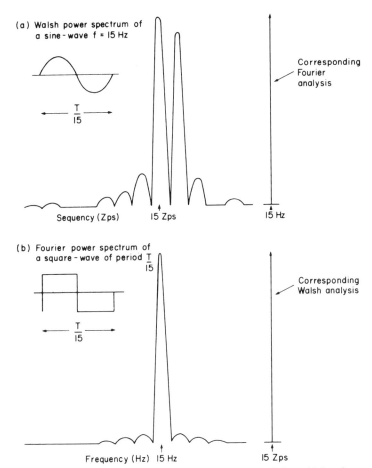

FIG. 3.3. Comparison between Walsh and Fourier spectra of sinusoidal and rectangular waveforms.

Only method (1) can be truly invariant with circular time shift, although for very many purposes the periodogram method will lead to quite acceptable results fully justifying the considerable reduction in computational effort obtained in this way compared with an equivalent Fourier calculation [16].

In order to understand these operations and to establish (where appropriate) relationships between the Walsh and Fourier spectrum derivation we need to make use of a number of relationships which are summarised in Table 3.1. These relate to a discrete time series x_i of length N and complete the comparison between Walsh and Fourier characteristics commenced in Table 2.2. Some of these were discussed or derived in earlier chapters; others will be considered here.

TABLE 3.1. Further characteristics of the discrete Walsh and Fourier function series.

	Walsh		Fourier
Orthogonal property	$\sum_{i=0}^{N-1} \text{WAL}(m, i) \text{WAL}(n, i)$ $= N$ for $n = m$ $= 0$ for $n \neq m$	Orthogonal property	$\sum_{i=0}^{N-1} \exp\left(j\frac{2\pi mi}{N}\right) \exp\left(j\frac{2\pi ni}{N}\right)$ $= N$ for $n = m$ $= 0$ for $n \neq m$
Symmetrical property	$\text{WAL}(n, i) = \text{WAL}(i, n)$	Symmetrical property	Not applicable
Shift theorem	Not applicable	Shift theorem	$\cos(n \mp m)i = \cos(ni)\cos(mi)$ $\pm \sin(ni)\sin(mi)$ $\sin(n \pm m)i = \sin(ni)\cos(mi)$ $\pm \cos(ni)\sin(mi)$
Multiplication theorem	$\text{WAL}(k, i)\text{WAL}(p, i) = \text{WAL}[(k \oplus p), i]$ $\text{CAL}(k, i)\text{CAL}(p, i) = \text{CAL}[(k \oplus p), i]$ $\text{CAL}(k, i)\text{CAL}(p, i) = \text{SAL}[p \oplus (k-1)) + 1, i]$ $\text{CAL}(k, i)\text{SAL}(p, i) = \text{SAL}[(k \oplus (p-1)) + 1, i]$ $\text{SAL}(k, i)\text{SAL}(p, i) = \text{CAL}[(k-1) \oplus (p-1), i]$ $\oplus$ = modulo-2 addition	Multiplication theorem	$2\cos ki \cos pi = \cos(k-p)i + \cos(k+p)i$ $2\sin ki \cos pi = \sin(k-p)i + \cos(k+p)i$ $2\cos ki \sin pi = -\sin(k-p)i + \sin(k+p)i$ $2\sin ki \sin pi = \cos(k-p)i - \cos(k+p)i$
Dyadic convolution $\circledast$	$x_i \circledast y_i = \frac{1}{N} \sum_{i=0}^{N-1} x_i y_{(\tau \oplus i)}$	Convolution *	$X_i * Y_i = \frac{1}{N} \sum_{i=0}^{N-1} x_i y_{(\tau + i)}$

τ = convolution delay

Convolution product relation	$x_i \circledast y_i \leftrightarrow X_k \cdot Y_k$	$\leftrightarrow$ = Transform	Convolution product relation	$x_i * y_i \leftrightarrow X_n \cdot Y_n$
Dyadic autocorrelation	$R_w(\tau) = \dfrac{1}{N}\sum\limits_{i=0}^{N-1} x_i x_{(\tau \oplus i)}$		Autocorrelation	$R_F(\tau) = \dfrac{1}{N}\sum\limits_{i=0}^{N-1} x_i x_{(\tau+i)}$
Dyadic spectrum	$P(0) = X_c^2(0)$ $P(k) = X_c^2(k) + X_s^2(k)$ $P\left(\dfrac{N}{2}\right) = X_s^2\left(\dfrac{N}{2}\right)$ $k = 1, 2, \ldots, \left(\dfrac{N}{2}-1\right)$		Fourier spectrum	$P_{(n)} = \dfrac{1}{N}\left|X_n\right|^2$
Dyadic Wiener–Khintchine theory Parseval's theorem	$R_w(\tau) \leftrightarrow P(k)$ $\dfrac{1}{N}\sum\limits_{i=0}^{N-1}[x_i]^2 = \sum\limits_{k=0}^{N-1}[X_k]^2$		Wiener–Khintchine theory Parseval's theorem	$R_F(\tau) \leftrightarrow P(n)$ $\dfrac{1}{N}\sum\limits_{i=0}^{N-1}|x_i|^2 = \sum\limits_{n=0}^{N-1}|X_n|^2$

3.3.1 The sequency spectrum via the autocorrelation function

The lack of an equivalent shift theorem for the Walsh function when this is considered in terms of spectral analysis means that the derivation of the power spectrum via the autocorrelation function (Wiener–Khintchine relationship), which represents the *classical method* of Fourier spectral analysis, is not directly possible.

However, it has been shown by Gibbs [1] that a dyadic equivalent of this relationship between autocorrelation and the power spectrum is valid, provided that the time shift in the correlation operation is obtained from modulo-2 addition.

Thus for

$$R_W(\tau) = \frac{1}{N} \sum_{i=0}^{N-1} x_i x_{\tau \oplus i} \tag{3.11}$$

we can write

$$R_W(\tau) \leftrightarrow P_W(k) \tag{3.12}$$

where $P_W(k)$ is the discrete Walsh power spectrum expressed in terms of a sequency series and $R_W(\tau)$ is the dyadic autocorrelation. This has become known as the logical derivation of the Wiener–Khintchine theorem.

A major advantage of deriving the Walsh power spectrum in this way is that the spectra obtained will be phase invariant. To obtain the spectrum by this route an additional calculation step is involved. This is shown in Fig. 3.4. The arithmetic autocorrelation is obtained first and the result transformed through a matrix operation to yield the dyadic autocorrelation of the function. This may then be transformed by using the FWT to obtain the power spectrum through Eq. (3.12). The additional step required is the correlation matrix operation needed to convert from the arithmetic to the dyadic correlation function. Whilst the Walsh transformation is obtained quite rapidly, the correlation matrix operation involves N additions, and if N is large this represents a serious lengthening of the calculations involved.

The translation between the correlation functions of these two systems has been shown by Pichler to take the form of two linear matrices

$$\mathbf{T}_{A-L} = \mathbf{D}_N \cdot \mathbf{T}_N \quad \text{and} \quad \mathbf{T}_{L-A} = \mathbf{T}_N^{-1} \cdot \mathbf{D}_N^{-1} \tag{3.13}$$

where $\mathbf{T}_{A-L}$ refers to translation from arithmetic to logical correlation, $\mathbf{T}_{L-A}$ to translation from logical to arithmetic correlation and $\mathbf{D}_N$ an $N \times N$ diagonal matrix whose elements are simply related to binary representation for the numbers 0, 1, 2, . . . , $N-1$. Here $\mathbf{T}_N$ is also an $N \times N$ matrix generated recursively [17].

3.3 Spectral Analysis

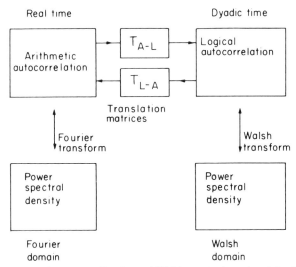

FIG. 3.4. Relationship between Fourier and Walsh spectra in real and dyadic time.

A recursive algorithm using a 'shuffling' matrix is defined by Robinson [18] and will permit the required matrix and its inverse to be developed for a known size of the series N, commencing with $\mathbf{T}_1^{-1} = 1$.

Gibbs and Pichler [19] have succeeded in combining the two steps into one matrix multiplication but at a cost of N^2 multiplications and additions; this is even slower than carrying out the translation and fast transform separately.

A more economical approach is carried out by Yuen [20], who has established a relationship between the Walsh power spectrum and the autocorrelation function without the intermediate calculation via the logical correlation function. To do this a square matrix consisting of $N \times N$ values operates on a column vector made up of sampled values of the autocorrelation function. This square matrix is related to a Paley-ordered matrix, and its rows obey certain recursive relations noted by Shanks [21] in his FWT derivation. Because of these relationships a measure of redundancy is found and the matrix may be factorised into a product of sparse matrices and multiplication carried out in a manner similar to the FWT. Yuen has shown that this power spectral calculation can be carried out in $3N \log_2 N$ additions plus $N \log_2 N$ divisions by 2. This latter simply represents a series of shifts to the right in a holding register.

3.3.2 The periodogram approach

The power spectrum coefficients can also be determined in a manner analogous to the periodogram used in Fourier power spectral analysis, viz.,

$$P_F(k) = \text{Re}(X_k)^2 + \text{Im}(X_k)^2 \tag{3.14}$$

where $\text{Re}(X_k)$ is the real (sinusoidal) component of the complex Fourier transform and $\text{Im}(X_k)$ the imaginary (cosine) component. Thus, for the Walsh spectrum we would calculate

$$\begin{aligned} P_W(0) &= X_c^2(0) \\ P_W(k) &= X_c^2(k) + X_s^2(k) \\ P_W(N/2) &= X_s^2(N-1) \end{aligned} \tag{3.15}$$

where $k = 1, 2, \ldots, (N/2) - 1$, giving $(N/2) + 1$ spectral points.

It should be noted that, whilst Eq. (3.15) can represent energy and hence conforms to Parseval's theorem, $\sqrt{P_W(k)}$ does not give a sequency amplitude spectrum as obtained with $\sqrt{\cos^2 \omega t + \sin^2 \omega t}$ in circular function theory since an addition theorem similar to that found with sine and cosine would be needed.

This derivation of the Walsh power spectrum is not, of course, completely invariant of phase shift of the input signal. It should be noted that Ulman's R sequency-ordered transform (Section 2.6) is invariant with phase shift and that its coefficients may be squared and normalised to yield a phase-invariant energy spectrum. A difficulty with the R transform is that it has a sequency-filtering action on the signal which reduces the value of the higher sequency terms, thus making the method of little value for other than strongly periodic data.

A complete review of the theory of the Walsh periodogram is given by Morettin [22]. He shows that the Walsh periodogram is, like its equivalent Fourier periodogram, not highly consistent, and that a more exact spectrum estimation is obtained for stationary signals by employing a smoothing operation based on signal partitioning. Here the signal is divided into L equal parts and a periodogram estimation is carried out for each part and the results averaged to give $[(N/2) + 1]/L$ spectral values.

An averaged Walsh power spectrum which is completely invariant to time shifts of the input data is derived by Dinstein and Silberberg [23]. It takes as its basis a periodogram spectrum definition similar to that given by Eq. (3.15) and proceeds to average these results for each of the $N/2$ spectra produced over all N possible distinct circular shifts. In their derivation it is proved that the average power spectral coefficients derived from transformed coefficients of the same sequency are equal. Hence it is only necessary to use the cosine transform coefficients in an expression for the averaged Walsh power spectrum, viz.,

$$P_{aw}(0) = X_c^2(0), \qquad P_{aw}(k) = \frac{1}{N} \sum_{m=0}^{N-1} X_c^2(k, m)$$

$$k = 1, 2, \ldots, (N/2) - 1, \qquad m = 0, 1, \ldots, N-1 \qquad (3.16)$$

For a deterministic periodic sequence having a period N, this power spectrum is the average of the Walsh power spectra of all N possible distinct circular shifts. What the method achieves is the averaging of all the first CAL squared spectral coefficients in a set of N periods of data sequences followed by the average of the second CAL squared coefficients and so on. It obtains this by setting up a series of N sets of the data, each obtained through a circular shift of the original data series x_i by m elements to the right. The average of each set of k spectral values through Eq. (3.16) gives an invariant spectrum since by incorporating all possible circular shifts of the signal, x_i in the average power spectrum $P_{aw}(k)$ must also include all possible sequency shifts in the sequence.

The motivation for the use of the averaged Walsh power spectrum lies in shape determination and pattern recognition, where the discontinuous nature of the Walsh functions enables a more compact spectra to be obtained [24]. The computation of Eq. (3.16) can be lengthy since it involves the derivation of a set of N transformations of an N-point series as well as the squaring and averaging process. However, if the derivation is carried out via the autocorrelation process as defined by Yuen [4] and use is made of the symmetry characteristic of the arithmetic autocorrelation function, where

$$R_{(N/2)-k} = R_{(N/2)+k} \qquad (3.17)$$

some reduction in the number of terms to be calculated is possible. A fast transformation from the arithmetic autocorrelation function to the averaged Walsh power spectrum making use of this concept is described by Dinstein and Silberberg [23].

3.3.3 The odd-harmonic sequency spectrum

The second definition for a sequency power spectrum introduced in Section 3.3 is based on the addition of the energy contents for groups of sequencies [25]. It results in a highly compressed spectrum of $p + 1$ points which is invariant to the cyclic shift of the input waveform. As with the periodogram spectrum, the resultant CAL and SAL functions are squared and summed. However, in this case they are grouped as the squared values of odd harmonics of a fundamental sequency before being added together. This is

Sequency² coefficient	Power spectral coefficient
CAL(0, t)	P_0
SAL(8, t)	P_1
SAL(4, t) CAL(4, t) } +	P_2
SAL(2, t) CAL(6, t) CAL(2, t) SAL(6, t) } +	P_3
SAL(1, t) CAL(7, t) CAL(3, t) SAL(5, t) CAL(1, t) SAL(7, t) SAL(3, t) CAL(5, t) } +	P_4

FIG. 3.5. The odd-harmonic sequency spectrum for $N = 16$. (P_0 = zero sequency coefficient; P_1 = eighth sequency coefficient; P_2 = fourth sequency coefficient; P_3 = odd-harmonics of second sequency; P_4 = odd-harmonics of first sequency.)

shown in Fig. 3.5 for $N = 16$. An inspection of this figure will show that the groupings follow that of natural order for the Walsh coefficients (see Table 1.7), and this enables an efficient in-place transform algorithm to be used.

The $p + 1$ components of the odd-harmonic sequency spectrum are defined as

$$P_0 = \text{HAD}^2(0, t), \qquad P_i = \sum_{k=2^{i-1}}^{z^i - 1} \text{HAD}^2(k, t)$$

$$i = 1, 2, \ldots, p, \qquad n = 0, 1, \ldots, N, \qquad p = \log_2 N \qquad (3.18)$$

It is not necessary to evaluate all the transform coefficients before carrying out the summation. A simplified signal flow diagram for the complete spectrum, suggested by Ohnsorg [14], is shown in Fig. 3.6. It will be seen that the sparse nature of this computation matrix permits considerable economy in the calculation of the spectral coefficients.

A similar power spectrum based on the Haar transform can also be defined. If we define the effective sequency of the Haar function series as 'one-half the average number of zero crossings per unit time interval', then it is seen that the Haar functions also fall into discrete groups, each member of a group having the same effective sequency as other members of the same group (Table 3.2). This was referred to earlier as the definition by order and

3.3 Spectral Analysis

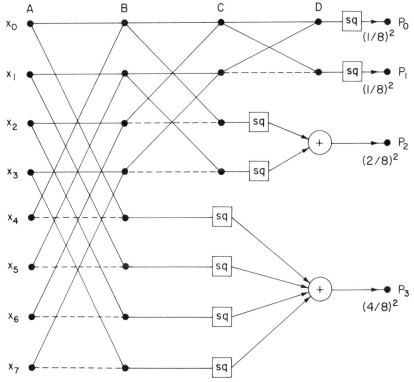

FIG. 3.6. Flow diagram for the odd-harmonic sequency power spectrum.

degree, given by Eq. (1.70). From this definition a power spectrum can be defined which can be considered as analogous to the sequency or frequency energy spectrum.

The procedure is to take the sum of the normalised value of the squares of the line spectra that fall within each grouping and use these values to indicate

TABLE 3.2. Haar sequency grouping, $N = 16$.

	Sequency group
HAR(0, t)	1
HAR(1, t)	2
HAR(2, t), HAR(3, t)	3
HAR(4, t), HAR(5, t) HAR(6, t), HAR(7, t)	4
HAR(8, t), HAR(9, t) HAR(10, t), HAR(11, t) HAR(12, t), HAR(13, t) HAR(14, t), HAR(15, t)	5

the energy contained in the Haar transformation. The resulting spectrum will be compressed somewhat differently than the odd-harmonic Walsh spectrum, but since it is computationally even simpler to obtain it may be used with advantage in certain applications [26]. The comparison between the two forms of coefficient groupings is commented upon by Ahmed *et al.* [27], who also point to similarities in the calculation of the two spectra. This leads to the consideration of a generalised C – T type of algorithm which can be used to compute either spectrum.

Both the Walsh and Haar compressed spectra suffer from the disadvantage of producing only p spectral coefficients for a data series set of 2^p points. As a consequence, they are not able to characterise the sampled data set very well. Despite this fact, its economy in computer calculation time is attractive and has resulted in a number of applications. Examples are found in systems analysis [28], image coding [29], vocoder-speech-signal coding [30] and speech synthesis [31].

3.3.4 Comparisons among Walsh and Fourier averaged spectra

A choice of Fourier or Walsh spectral estimation for a given purpose will be found to be strongly dependent on the signal characteristics, as indicated earlier in connection with transform theory. Some examples will now be given to illustrate this dependence.

Figure 3.7 illustrates the Walsh and Fourier power spectra for a short-term transient signal obtained from a shock-excited mechanical structure. We can see from this figure that close similarities exist in the main region of the power spectra. However, a significant region of higher sequency power is present in the Walsh case, well removed from the main region found in both representations. The underlying reasons for this are associated with the origin of the particular signal being analysed. Since the signal is obtained from harmonic motion of a mechanical structure and hence defined by means of a linear differential equation, we can expect it to be represented by an exponential series, i.e., a sum of sinusoidal and cosinusoidal terms. We saw in Fig. 3.2 that a smoothly-varying sinusoidal signal will give rise to just these additional energy regions of higher-sequency terms when analysed in this way.

A similar region of high-order frequency power coefficients is found with the Fourier analysis of a rectangular synthesised signal, such as the comparative spectra of the pulse-coded modulation (PCM) waveform shown in the second example, Fig. 3.8. The Walsh spectral representation of the PCM signal shows a precise sequency-limited bandwidth due to the finite number of terms required to synthesise a binary-coded signal having a length which is

3.3 Spectral Analysis

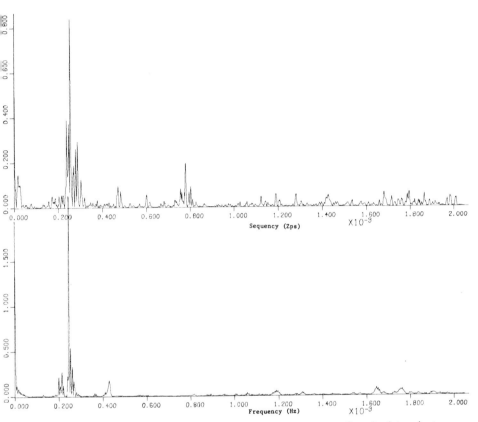

FIG. 3.7. Comparison between sequency and frequency power spectra for a shock transient.

related to the Walsh time base. In the Fourier case a theoretically unlimited series of harmonic coefficients is produced by the spectral analysis.

An example which exploits this difference in representation is shown in Fig. 3.9, which concerns the differences in the Walsh power spectra obtained for two rectangular types of signal. The first signal (a) is derived from a given Morse code message of 15 characters which is digitally sampled into 512 samples having either of two values, 0 or 1. The mark/space ratio of the encoded morse is maintained at a precise value, such as would be obtained with a machine-generated code. The second signal (b) represents the same message, but this time the mark/space ratio is varied slightly in a random manner, as would be experienced, for example, if the message were sent with a handkey. Comparison between the two spectra shows very clearly the essential difference between the two sets of code. A regular sequency-limited spectrum is obtained in the machine-sent case, showing that discrimination

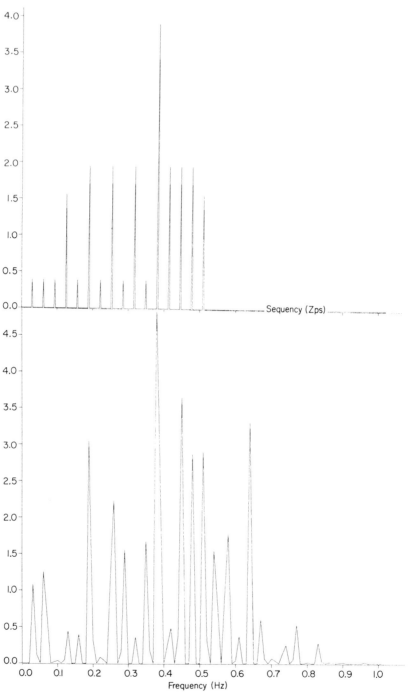

FIG. 3.8. Comparison between the sequency and frequency spectra of a PCM waveform signal.

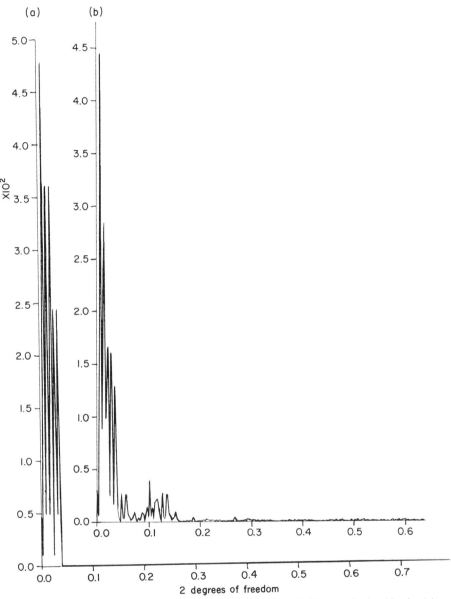

FIG. 3.9. Comparison between the Walsh transformation of a Morse code signal having (a) uniform mark/space ratio and (b) random mark/space ratio.

between the two cases can easily be obtained in this domain whereas the difference would be difficult to detect in the original time domain.

These examples indicate clearly the respective roles of Walsh and Fourier spectral analysis for discontinuous and smoothly-varying signals, respectively. Where the signal is derived from a sinusoidally-based waveform, such as would be obtained from a spring-mass-damped system (mechanical structure), Fourier analysis is relevant. Where the signal contains sharp discontinuities and a limited number of levels such that it may be synthesised from a combination of rectangular waveforms, Walsh analysis is appropriate. However, for some special purposes (e.g., on-line applications), the speed with which the Walsh transform may be obtained by digital computation becomes more important in the analysis of sinusoidally-based signals despite the increased complexity in the spectrum so obtained.

3.4 Digital filtering

The use of sequency functions for digital filtering allows a simplification in the modelling of the filter transfer characteristic. Whereas with Fourier functions any attempt to realise an ideal filter characteristic having unity gain inside the pass band and zero outside of it will result in the generation of unwanted perturbations (the Gibbs phenomenon), no such difficulty is found with sequency functions since these latter are well suited to jump discontinuities. Hence the need to taper or modify the transfer characteristic to achieve an acceptable compromise between transient response and sharp cut-off property can be avoided. As we shall see later, it is possible under some circumstances to achieve an ideal *brick-wall* filter by using sequency transformation methods.

Another reason for using Walsh and other sequency functions is, of course, the economy in the number of arithmetic operations required and the absence of complex arithmetic. This is particularly important for large-scale filtering tasks, e.g., image processing.

Early design for sequency filters was concerned with analog circuits in which the essential ingredients are integrators, switches and sample-and-hold amplifiers. Harmuth [32, 33] gives several examples of these Golden and James [34] have also described a resonant $L-C$ filter consisting of inductors, capacitors and a switch driven by a sampled Walsh function. Later developments of the resonant filter are the digital designs of Nagle [35] and Vandivere [36].

Digital filter development has used the fast sequency transformations either as a means of simplifying the hardware logic requirements or to produce efficient software filters. The lines of development for software filters have been constrained by the difficulty in carrying out convolution with these series and has resulted in the use of a generalised Wiener technique which is considered here as a major tool in sequency filtering applications. This may also be applied to two-dimensional filtering requirements and finds application in image transmission and analysis.

The basis for a number of these developments will be discussed next as a preliminary to a description of applications to be given in later chapters.

3.4.1 Generalised Wiener filtering

The familiar route to non-recursive digital filtering is via the inverse transformation of the product of the transforms of the signal and an impulse response function specifying the required filter performance [37]. In Fourier analysis this will give a result equivalent to the correlation of the signal and the impulse response function, which is what we require for a filtering action. When using sequency functions we have to do this somewhat differently due to the absence of a direct relationship between arithmetic correlation and transformed products, which was noted earlier in this chapter.

The technique used is a modification of the classical technique of Wiener filtering [38] which uses a *unitary transform matrix* and sampled data and is known as *generalised Wiener filtering* [39]. A matrix is said to be unitary if its inverse is its conjugated transposed version; this leads to ease in many computations involving such matrices [10]. The Hadamard, Walsh, Haar and Fourier (complex) matrices are all unitary.

Figure 3.10 shows the generalised one-dimensional Wiener filtering system. A signal vector x_i is assumed to consist of additive zero-mean signal s_i and noise n_i components which are assumed to be uncorrelated with each other. A unitary transformation operation utilising an $N \times N$ matrix $\mathbf{A}$ is performed on x_i to yield

$$\mathbf{X}_k = \mathbf{A} x_i = \mathbf{A} s_i + \mathbf{A} n_i = \mathbf{S}_k + \mathbf{N}_k \qquad (3.19)$$

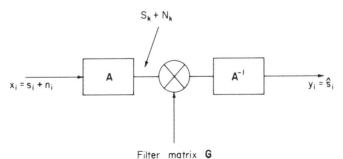

Fig. 3.10. Generalised Wiener filtering.

The resultant vector $\mathbf{X}_k$ is multiplied by an $N \times N$ filter matrix $\mathbf{G}$ and inversely transformed to produce a filtered output

$$y_i = \mathbf{A}^{-1} \cdot \mathbf{G} \cdot \mathbf{A} \cdot x_i \qquad (3.20)$$

This is the familiar transform–modify–inverse transform method of filtering. If $\mathbf{G}$ is chosen correctly, then the required filtered output will consist of the signal component s_i plus a much reduced noise component n_i.

Pratt [39] has shown that discrete Wiener filtering may be implemented by any unitary transformation, including Fourier, Walsh, Haar and Karhunen–Loève, for the same MSE. Fourier non-recursive filtering is a special case of Wiener filtering in which the filter matrix is a vector representing a conjugate symmetric set of filter weights derived from sampling the required frequency response. Thus we are free to use any orthogonal transformation that will minimise the computational processes involved in filter generation and operation.

Filter design is concerned primarily with the choice of the filter matrix $\mathbf{G}$, which is chosen in conjunction with the transform to minimise the MSE between the required signal s_i and its estimated value $\hat{s}_i = y_i$. The characteristics of the $N \times N$ filter matrix $\mathbf{G}$ will have a pronounced effect on the efficiency of the filtering process. In the optimum case the filter matrix will have a number of non-zero coefficients off the diagonal axis, and N^2 multiplications with the transformed signal will be required. This is known as *vector filtering*.

Since X_k will normally be calculated by using a fast transform requiring $N \log_2 N$ or fewer operations, then multiplication by the N^2 filter coefficients represents a major task. However, as with fast transform matrix calculation, it is possible to design $\mathbf{G}$ such that it contains few, if any, non-zero coefficients off the diagonal axis. With a matrix containing many zero terms, advantage can be taken of the redundancy obtained to devise fast multiplication techniques.

In very many cases it is possible to arrange for $\mathbf{G}$ to contain only non-zero diagonal elements so that the filtering operation becomes simply the weighting of each transformed coefficient of X_i individually and only N multiplications are required. This sub-optimal case is known as *scalar filtering*.

In the design of the generalised Wiener filtering system of Fig. 3.10, the filter matrix $\mathbf{G}$ should be chosen to minimise the mean-square error

$$\text{MSE} = \sum_D \overline{(s_i - \hat{s}_i)(s_i - \hat{s}_i)'} \qquad (3.21)$$

where Σ_D is the sum of elements along the main diagonal, $(s_i - \hat{s}_i)$ equal values arranged as a vertical vector, $(s_i - \hat{s}_i)'$ equal values arranged as a horizontal vector and an overbar indicates an averaging process.

3.4 Digital Filtering

It can be shown that this may be expressed in terms of transform domain quantities as

$$\text{MSE} = \sum_D (C_S - 2GC_S + GG(C_S + C_N)) \quad (3.22)$$

where C_S and C_N are the covariance matrices of the signal and noise components of x_i given in Eq. (3.19). Determination of the least MSE from Eq. (3.22) yields the optimum filter matrix

$$G_0 = C_S/(C_S + C_N) \quad (3.23)$$

and since C_S and C_N are related to the power spectral densities of the signal and noise, it is possible to find **G** if we know the spectral form of the required and the noisy signals.

The filter matrix G_0, of course, applies to vector filtering and will have finite coefficient values off the diagonal. Pearl [13] has considered applying a constraint to the value of **G** obtained such that all off-diagonal elements are zero. The improvement obtained in signal-to-noise ratio of the filtered signal is sufficient in most cases to implement scalar filtering in this way.

Alternatives considered by Pratt [39] are to select only the diagonal and near-diagonal element of G_0, making the remaining elements zero, with the location of the zero elements such as to enable fast computational algorithms to be employed. Another technique is to select only those elements of large value (threshold selection).

Advantage may be taken of the spectral density relationships implied in Eq. (3.23) to devise a procedure for derivation of **G** where the form of the desired output signal is known. This is described by Rumatowski and Sawicki [40] for a scalar filter employing either Walsh or Haar transformation, where

$$G_i = X_S(i)/X_{S+N}(i), \quad i = 0, 1, \ldots, N \quad (3.24)$$

where X_S is the Walsh or Haar transformation of the desired output signal s_0 and X_{S+N} a similar transformation of the input signal s_i containing s_0. Coefficients for the diagonal coefficients for **G** are taken as the ratio of the ith elements of X_S and X_{S+N} with $G_i = 0/0$ (where it occurs) being taken as 0.

The number of real operations required by this method is $4N \log_2 N$ additions/subtractions plus N multiplications in the case of the Walsh transform and $4(N-1)$ real additions plus N multiplications for the Haar transform. Since this method ignores the off-diagonal coefficients for G_i, some error is inevitable, although error can be absent in some cases.

A similar scalar filter employing a hybrid Walsh–Haar transformation has been described by Rao et al. [41].

3.4.2 Threshold filtering

The simplest form of Wiener filtering occurs when the scalar filter matrix is limited to values 0 and 1. In the case of a low-pass filter, for example, all the sequency coefficients above a certain value are set to zero through multiplication by the zero coefficients of G_i before retransformation takes place. This is an extremely efficient process and can result in a perfect brick-wall filter.

Analog signals are handled by such a filter through conversion to a stepwise approximation to the signal. The signal is integrated over the step length and held at its final value by a sample-and-holder amplifier. Since the stepped approximation obtained cannot contain any Walsh function components of higher sequency than the step rate, an ideal (sequency) low-pass filter results. An example is given in Fig. 3.11, in which a low-sequency signal (a) has added to it a second signal (b) comprised of elements of higher sequency. Carrying out the Wiener process by using a series of ones in the top of the filter matrix and zero elsewhere results in the original low-sequency signal being reconstituted exactly. The process is ideally suited to filtering of rectangular data to which a noise component is added, such as would occur, for example, in digital transmission systems.

A recursive form of threshold filtering is described by Eghbali [42]. A sequency-ordered Walsh transformation is carried out in the image matrix, and initially the highest sequency coefficients along the Nth row and Nth column are set to zero. The result is retransformed to form a filtered image, and the average MSE between the original and retransformed image is calculated. The next-highest sequency coefficients are then set to zero, and after retransformation a second error figure is calculated between the result and the previously derived image. Recursion proceeds if this error shows an improvement over the preceding error figure; otherwise the procedure is halted and the preceding image retained as the final filtered output. The method is equally applicable to a single-dimensional series. This recursive technique has been applied to the transformed and filtered data in a scalar Wiener filtering system prior to the retransformation stage and shows some improvement over the Wiener filtered output.

FIG. 3.11. Example of sequency-limited filtering.

A development of simple threshold filtering based on the known properties of the original signal can result in an efficient matched filtering technique which may be used to recover a known signal immersed in noise. Here the form of the signal is known, so that some distortion of the recovered signal is acceptable.

The relationship between the transformed values of the matching signal and the set of limited value filter weights is obtained by first defining a threshold level in the magnitude of the transformed series. A corresponding series of weights is formed in which all values in excess of this threshold value are made 1 and all these below it are made 0.

If we consider the synthesis of a function by means of the summation of a number of Walsh functions, it becomes apparent that the threshold level must be set as to allow sufficient of these functions to be summed to form an acceptable reconstruction of the filtered signal. Thus, for smoothly-varying matching signals we would expect a low threshold level to minimise distortion in the reconstructed signal. Where only a few terms are required to synthesise the signal adequately, such as pulse or discontinuous functions, a high threshold level can be used.

A flow diagram for a simple matched filtering program is given in Fig. 3.12. The threshold level L is defined as a percentage of the largest sequency coefficient value found in the Walsh transform of the matching signal. The threshold level is provided as a parameter to the program together with the number of samples (N) for the signal and matching signal.

Figure 3.13 gives an example concerning a series of pulses obscured by random noise. At a given level of $L = 33\%$, the locations of the pulses are clearly determined, although some distortion of their relative height and width is seen. The method has been applied successfully to the recovery of geological data [43] and in the enhancement of video images [44] (see also Brown [45]).

3.4.3 Frequency-based scalar filtering

A difficulty in the implementation of general frequency filters based on the use of sequency functions and employing generalised Wiener filtering is to obtain a suitable method for determining the filter weights. In the preceding examples either the form of output is known or the requirement is simply to improve the signal-to-noise ratio. There is, however, a large amount of filter design information available for filtering sampled data via the Fourier transform and, somewhat naturally, attempts have been made to utilise this through a relationship between the Fourier and Walsh Wiener filtering equations.

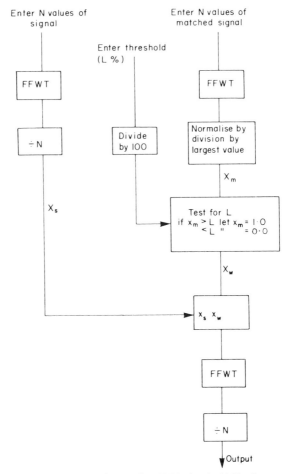

FIG. 3.12. Flow diagram for Walsh threshold filtering.

Kahveci and Hall [46] have shown that, if we express the two forms of filtering in matrix terms, we can obtain this relationship. We can write the output column vector $y_1(i)$ for Fourier filtering as

$$y_1(i) = \mathbf{F}^{-1}\mathbf{G}_1\mathbf{F} \cdot \mathbf{x}_i \tag{3.25}$$

where $\mathbf{F}$ and $\mathbf{F}^{-1}$ are the direct and inverse Fourier transform matrices and $\mathbf{G}_1$ is the set of filter weights. Walsh filtering can be written in the same way as

$$y_2(i) = \mathbf{W}^{-1}\mathbf{G}_2\mathbf{W} \cdot \mathbf{x}_i \tag{3.26}$$

where $\mathbf{W} = \mathbf{W}^{-1}$ for the direct and inverse Walsh transform matrices and $\mathbf{G}_2$ is a second set of filter weights appropriate to Walsh filtering. If we assume

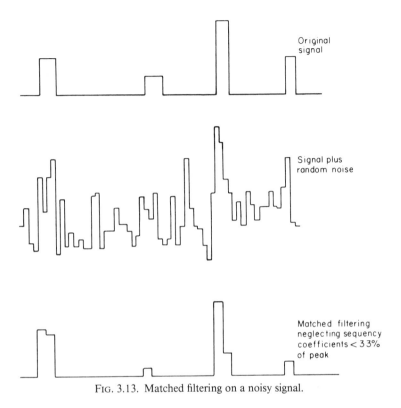

FIG. 3.13. Matched filtering on a noisy signal.

similar outputs for the two filtering operations, then $y_1(i) = y_2(i)$ and we can write

$$F^{-1}G_1F \cdot x_i = WG_2W \cdot x_i \qquad (3.27)$$

from which

$$G_2 = WF^{-1}G_1FW \qquad (3.28)$$

The filter weights G_1 may be derived by using well-tried methods. By using the fast Fourier and Walsh transforms, the required filter weights G_2 necessary for Walsh scalar filtering are obtained. The G_2 matrix obtained may not necessarily be diagonal but can be transformed into a diagonal matrix by row or column transformations. Whilst this process of filter weight derivation will be slower than simple determination of G_1 for Fourier filtering, the actual productive process of filtering using the Walsh transform can be considerably faster. This would be important for complex or repetitive filtering operations such as two-dimensional image filtering or on-line television applications.

In the case of a single filtering operation on N samples, a maximum of

$2N \log_2 N$ additions/subtractions plus N^2 real products will be required. The calculation involving the N^2 real products will generally be the dominant factor in determining the speed of the filtering operation and may be reduced by a process of selective computation, as originally suggested by Pratt et al. [29]. In this case only the non-zero elements are subject to matrix multiplication, thus reducing the number of mathematical operations needed. An improvement of up to 75% reduction in computational time has been claimed for image-enhancement applications.

3.4.4 Signal-to-noise ratio improvement

An important filtering activity is one directed towards the improvement in the signal-to-noise ratio for a noisy signal. Several sequency techniques have been proposed and have been found particularly successful where the signal takes the form of a rectangular pulse such as we find in radar applications.

A transformation technique using the Walsh–Hadamard transform is described by Thompson and Rao [47]. It is applicable where the transmission system has unused capacity. For example, in the case of a particular coding which permits three out of the eight used coded data sequences to be redundant (see Chapter 7), the noise associated with these extra channels can be removed by setting the three transformed coefficients at zero. Since this is carried out in the transformed domain, where the noise arising from all the channels is evenly distributed over the spectrum, a signal-to-noise improvement is achieved upon retransformation. This can be accomplished through a transform algorithm in which the unneeded signal flow paths are omitted. The selection of sequencies for reduction to zero value is dependent on the filtering characteristics required. Three cases of hard limiting, whitening and Gaussian noise characteristics are given.

Filtering can also be arranged through time domain averaging. An economical method exploiting the nature of the Haar transform is proposed by Karpola and Jernigan [48]. Several techniques are suggested. Where the period of the noisy signal is known, a Haar transformation may be carried out for each repetition of the signal occupying a given period. The small number of transformed coefficients obtained with each signal period are averaged with corresponding Haar coefficients in each transformation. Inverse transformation of the averaged set of coefficients recovers the signal, with an improvement in the signal-to-noise ratio. This method of sequency domain averaging has the advantage over time domain averaging of requiring considerably fewer averaging operations. It is also possible to carry out

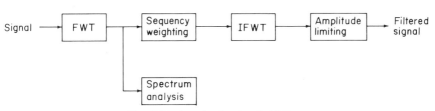

FIG. 3.14. An adaptive threshold filter.

sequency threshold filtering before averaging, thus reducing the number of operations required still further.

An alternative method is to perform first a *perfect shuffle* on the entire data sequence (see Subsection 4.2.1). The perfect shuffle has the effect of arranging all the first samples from each period sequentially, followed by all the second samples from each period, and so on. For a signal consisting of p periods of a known signal and $N = 2^k$ samples per period, it is clear that for a deterministic signal the first p shuffled values will be identical, as will the next group of p values and all subsequent groups. It can be shown that, since a Haar basis function which spans intervals equal to or less than p points will contain zero-valued coefficients, a Haar transformation of the complete sequence of $T = pN$ samples will only have N non-zero coefficients; the first N values. Thus for a signal containing Gaussian noise much of this will be removed if the remaining $T - N$ transformed values are made zero before retransformation is carried out to recover the filtered signal. This method allows an improvement in signal-to-noise ratio without explicit averaging. The improvement is in all cases proportional to the number of periods averaged. The combination of Haar averaging plus threshold filtering has been shown to give an increase in signal-to-noise ratio up to 31 dB for 32 periods of the deterministic signal [48]. An adaptive threshold scheme is described by van Cleave [49, 50] who uses the Walsh transform in a fast real-time hardware filter. This processes the noisy output of a microwave receiver (e.g., passive radar or a surveillance system). The signal is subject to an FWT operation, and some selective sequency limitation of the transformed signal takes place before the modified signal is transformed back into the time domain. A thresholding operation removes some of the remaining noise products contributing to the overall signal-to-noise ratio improvement for the pulse output of the system. The sequency limiting described is an adaptive one based on measurements on the sequency domain signal. A signal analysis operates on the known difference in the sequency spectrum between the noise-only condition and the noise-plus-pulse condition to produce threshold values for sequency filtering (Fig. 3.14). A Gaussian amplitude probability model is taken for the noise-only case. An improvement of about 20 dB is claimed for the hardware system operating at a rate of 20 MHz.

Descriptions of other signal-to-noise improvement techniques can be found in Chapters 5 and 6.

3.4.5 Two-dimensional filtering

The generalised Wiener filtering method can be extended readily into two dimensions. If we consider

$$x_{ij} = s_{ij} + n_{ij} \tag{3.29}$$

to represent an $N \times N$ matrix of sampled digital data comprising a signal s_{ij} and noise n_{ij} (e.g., a digitised image containing unwanted aberrant elements), then the Wiener filtering operation can be expressed as

$$\mathbf{X}_{m',n'} = \sum_{m=0}^{N-1} \sum_{n=0}^{N-1} \mathbf{X}_{m,n} \mathbf{G}(m, m', n, n') \tag{3.30}$$

where $\mathbf{X}_{m,n}$ is the two-dimensional transform of x_{ij} and $\mathbf{G}(m, m', n, n')$ is a weighting function [51].

To derive each element of the filtered data in the transform domain $\mathbf{X}_{m',n'}$, an N^2 product–sum is required. A further N^2 product–sum is needed to reconstruct the filtered image by means of the corresponding inverse transformation

$$\hat{s}_{ij} = \sum_{m'=0}^{N-1} \sum_{n'=0}^{N-1} \mathbf{X}_{m',n'} \mathbf{K}(i, j, m', n') \tag{3.31}$$

where $\mathbf{K}$ is the kernel of the inverse transformation used.

It is usually possible, however, to separate the two-dimensional transformation into sequential operations on rows and columns of the array, thus simplifying Eqs. (3.30) and (3.31) to

$$\mathbf{X}_{m',n'} = [\mathbf{G}(n, n')][\mathbf{X}_{m,n}][\mathbf{G}(m, m')] \tag{3.32}$$

and

$$\hat{s}_{ij} = [\mathbf{K}_{i,n'}][\mathbf{X}_{m',n'}][\mathbf{K}_{j,m'}] \tag{3.33}$$

where the bracketed terms indicate matrix representations of the appropriate arrays. Here $\mathbf{G}(n, n')$ and $\mathbf{G}(m, m')$ are the Wiener filtering matrices for the rows and columns, respectively, of the original data array. Under these conditions two-dimensional Wiener filtering becomes the sequential application of one-dimensional filtering applied to rows and columns of the input data array.

Optimum filter design will generally result in filter coefficients which have levels other than zero or one. It is possible to simplify the practical application of such a filter by replacing those filter coefficients having very small values, close to zero, with zero values. Pratt [52] has claimed that, under certain conditions, up to 90% of the filtering multiplications can be avoided in this way.

Non-linear filtering involves carrying out a non-linear operation on each of the transformed samples. Two types of non-linear operation that have been employed in the area of image processing are logarithmic operation

$$X_{m',n'} = \frac{KX_{m,n} \log\{|X_{m,n}|\}}{|X_{m,n}|} \qquad (3.34)$$

and a power-law operation

$$X_{m',n'} = \frac{X_{m,n}|X_{m,n}|^k}{|X_{m,n}|} \qquad (3.35)$$

where K and k are constants.

A logarithmic operation will attenuate transform domain samples by a factor proportional to their magnitude and can give an edge enhancement to a processed picture. It is related to a similar Fourier non-linear operation known as the *cepstrum*, which has been used successfully to remove cyclic background noise [53].

A power-law operation tends to emphasise the difference between low- and high-amplitude samples and can also be used for image enhancement. A number of examples of picture filtering using these techniques are given by Pratt [52].

Two-dimensional filtering through dyadic convolution has been suggested by Rao [54]. It also uses a single-dimensional model translated into two dimensions and realised through a dyadic convolution filter. This essentially carries out the well-known transfer function equation

$$Y = XS \qquad (3.36)$$

where Y and X are the transformed output and input data and S is the transform of the filter system response function. Implementation of dyadic convolution is, however, hampered by lack of a suitable fast computational method and the difficulty of defining the system response in sequency terms (see Subsection 3.2.1).

3.5 Waveform synthesis

The traditional method of representing arbitrary integrable time functions is by superposition of a number of sine or cosine functions. Similar results can be obtained by using sequency functions, and due to their orthogonal characteristics both the Walsh and Haar series have been used for this purpose. The advantages obtained by the use of these functions lies in their ease of generation and rectangular format, which make them easy to manipulate in digital systems, particularly with the hardware systems which are described in Chapter 4.

If we consider first the synthesis of an arbitrary function using a Walsh series, we note that theoretically an infinite series of terms is needed. In a practical situation of a truncated Walsh expansion of the function when not all of the terms required for exact representation are available, the truncated Walsh expansion can be shown [55] to have minimal integral squared error, viz.,

$$\text{MSE} = \frac{1}{N} \sum_{i=0}^{N-1} [x_i - y_i]^2 \qquad (3.37)$$

where x_i is the ith sampled value of the original time function $f(t)$ and y_i the synthesised ith sampled value of $f(t)$ using the truncated Walsh transformation

$$y_i = \sum_{n=0}^{N-1} X_n \, \text{WAL}(n, i) \qquad (3.38)$$

where $i = 0, 1, 2, \ldots, N-1$ and X_n is the coefficient of the Walsh function $\text{WAL}(n, i)$.

The MSE will approach zero as N becomes very large. It is not necessary to use a large value, however, to obtain reasonable accuracy. The procedure is to select a set of M dominant terms out of a complete set of N terms and to synthesise the waveform from this smaller set. It has been shown by Kitai (e.g., [56]) that a Walsh expansion of a periodic sinusoid can be made with only eight terms and that synthesis from this small number of terms will yield a wave whose only non-zero harmonics are the 31st and 33rd, the 63rd and 65th, etc., harmonic magnitudes being inversely proportional to frequency.

3.5.1 The dominant-term concept

The procedure described above is termed the *dominant-term concept*. In a synthesis described by Chen and Sun [57] an N-point DFT of a number of arbitrary waveforms is carried out and the first M coefficients selected from the set of Walsh coefficients arranged in descending order of absolute magnitude. Synthesis from these M values is described in terms of the expression

$$x_i = \sum_{k=1}^{M} X(n_k) \, \text{WAL}(n_k, i) \qquad (3.39)$$

where $i = 0, 1, \ldots, N$ and $k = 0, 1, \ldots, M$ and $X(n_k)$ is the kth most dominant Walsh coefficient amongst the first N coefficients. This is known as an approximate discrete inverse Walsh transform.

Synthesis of a commonly required test waveforms to an acceptable degree of accuracy can be achieved by choosing only 8 out of a possible 64 functions. Thus for a triangular test waveform the functions selected for x_i can be shown to be

$$x_i = 32 \text{ WAL}(0, i) - \text{WAL}(1, i) - 16 \text{ WAL}(2, i) - 8 \text{ WAL}(6, i)$$
$$- 4 \text{ WAL}(14, i) - 2 \text{ WAL}(30, i) - \text{WAL}(62, i) \qquad (3.40)$$

and for a sinusoidal waveform

$$x_i = 41 \text{ WAL}(1, i) - 2 \text{ WAL}(2, i) - 17 \text{ WAL}(5, i) - 3 \text{ WAL}(9, i)$$
$$- 8 \text{ WAL}(13, i) - 2 \text{ WAL}(25, i) - 4 \text{ WAL}(29, i) - 2 \text{ WAL}(61, i)$$
$$(3.41)$$

For these and other similar test waveforms MSE accuracies of between 0.2 and 2% have been achieved [57, 58].

The dominant term concept is also of value in speech synthesis and in transmission data compression applications. These topics will be considered in later chapters.

3.5.2 Effect of waveform characteristic

The synthesised periodic waveform will form a stepped approximation to the required arbitrary waveform and will, in fact, contain N steps per period in its reconstruction. The effect of this may be minimised by passing the result through a low-pass filter, or a higher sampling rate may be chosen.

As will be apparent from earlier discussions, the effective ratio M/N for acceptable MSE will depend on the characteristics of the waveform being synthesised. This may be demonstrated by carrying out the following procedure [59].

The waveform is transformed and normalised to give unit value for the largest frequency or sequency coefficient. A *threshold criteria R* is chosen to be less than 1 and all coefficients found to have value less than R reduced to 0. This enables a limited number of coefficients to be made available from which reconstruction of the original series is carried out. The accuracy of synthesis using these coefficients can then be compared for the Walsh and Fourier function series.

Typical examples of this approach are shown in Figs. 3.15 and 3.16 for $N = 1024$. The first diagram shows the effect of transformation, threshold-

132   3 Analysis and Processing

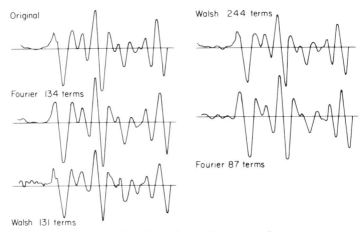

FIG. 3.15. Synthesis of a continuous waveform.

ing and reconstruction for a section of a continuous seismic waveform. Approximately twice the number of Walsh terms are required to give a similar accuracy as may be obtained in the Fourier case.

The second example shows the transformation and reconstruction of a rectangular waveform. This matches the form of the Walsh function and results in efficient reconstruction for considerably fewer Walsh terms than Fourier terms.

It will be recognised from these examples that a continuous type of waveform favours the Fourier transform whereas a rectangular waveform, or more precisely a discontinuous waveform, is reconstructed more easily with the Walsh transform. This would also be true for a comparison between the Fourier and Haar transformations.

3.5.3 Synthesis with Haar functions

The convergence features of an expansion in Haar functions are actually superior to the Walsh functions, as noted by Alexits [60], so that we would expect a small number of terms to be involved in any synthesis of arbitrary waveforms. This is certainly the case where we take the Haar terms to represent the number of periods of double inversion, which typifies the shape of the Haar functions, that may be contained in a given truncated field length. Thus to synthesise a 128-sample element function we need only take 7 discrete Haar series $F_1, F_2, F_4, F_8, F_{16}, F_{32}$ and F_{64}, where the subscripts of each variable is indicative of the number of cycles of double inversion for a given Haar component per data field length of 128 elements [61].

There is, however, a difficulty in binary representation and generation of

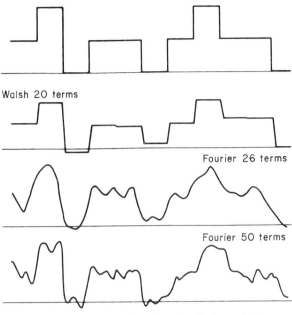

Signal reconstruction from limited number of Fourier and Walsh coefficients

FIG. 3.16. Synthesis of a discontinuous waveform.

the three-valued Haar function compared with the two-valued Walsh function, so that the Haar series is of less value in waveform synthesis, particularly in the design of hardware systems.

References

1. Gibbs, J. E. (1966). Walsh spectrometry: A form of spectral analysis well-suited to binary digital computation. National Physical Laboratory Report, Teddington, England.
2. Kennett, B. L. N. (1971). Introduction to the finite Walsh transform and the theory of the fast Walsh transform. *Proc. Theory Walsh Functions, Hatfield Polytechnic, England.*
3. Yuen, C. K. (1977). The auto and cross-correlations of Walsh functions. *SIAM J. Appl. Math.* **32**, 177–190.
4. Yuen, C. K. (1975). An algorithm for computing the correlation functions of Walsh functions. *IEEE Trans. Electromag. Compat.* **EMC-17**, 177–180.
5. Harmuth, H. F. (1972). "Transmission of Information by Orthogonal Functions." Springer-Verlag, Berlin.
6. Gordon, J. A., and Barrett, R. (1971). Digital majority logic multiplexer using Walsh functions. *IEEE Trans. Electromag. Compat.* **EMC-13**, 171–176.
7. Pitassi, D. A. (1971). Fast convolution using Walsh transforms. *Proc. Symp. Applic. Walsh Functions, Washington, D.C.*, AD727000, pp. 130–133.

8. Gibbs, J. E., and Gebbie, H. A. (1969). Application of Walsh functions to transform spectroscopy. *Nature* **222**, 1012–1013.
9. Frank, T. H. (1971). Implementation of dyadic correlation. *Proc. Symp. Applic. Walsh Functions, Washington, D.C.*, AD727000, 111–117.
10. Maqusi, M. (1981). "Applied Walsh Analysis." Heyden, London.
11. Polyak, B. T., and Schreider, Y. A. (1962). The application of Walsh functions in approximate calculations. *Voprosy Theor. Matem. Mashin Coll.* **II**, 74–90.
12. Gibbs, J. E., and Millard, M. J. (1969). Walsh functions as solutions of a logical differential equation. National Physical Laboratory Report 1, Teddington, England.
13. Pearl, J. (1971). Walsh processing of random signals. *IEEE Trans. Electromag. Compat.* **EMC-13**, 137–141.
14. Ohnsorg, F. R. (1971). Spectral modes of the Walsh–Hadamard transform. *Proc. Symp. Applic. Walsh Functions, Washington, D.C.*, AD727000, pp. 55–59.
15. Ahmed, N., Rao, K. R., and Schultz, R. B. (1971). The generalised transform. *Proc. Symp. Applic. Walsh Functions, Washington, D.C.*, AD727000, pp. 60–67.
16. Beauchamp, K. G. (1972). The Walsh power spectrum. *Proc. NEC* **27**, 377–382.
17. Pichler, F. R. (1970). Some aspects of a theory of correlation with respect to Walsh harmonic analysis. Depart. Electrical Engineering, Univ. of Maryland, Report R-70-71, College Park, Maryland.
18. Robinson, G. S. (1972). Discrete Walsh and Fourier power spectra. *Proc. Symp. Applic. Walsh Functions, Washington, D.C.*, AD744650, pp. 298–309.
19. Gibbs, J. E., and Pichler, F. R. (1971). Comments on the transformation of Fourier power spectra into Walsh power spectra. *IEEE Trans. Electromag. Compat.* **EMC-13**, 51–54.
20. Yuen, C. K. (1973). A fast algorithm for computing Walsh power spectrum. *Proc. Symp. Applic. Walsh Functions, Washington, D.C.*, AD763000, pp. 279–283.
21. Shanks, J. L. (1969). Computation of the fast Walsh Fourier transform. *IEEE Trans. Comput.* **C-18**, 457–459.
22. Morettin, P. A. (1981). Walsh spectral analysis. *SIAM Review* **23**, 279–290.
23. Dinstein, I., and Silberberg, T. (1981). Average Walsh power spectrum for periodic signals. *IEEE Trans. Electromag. Compat.* **EMC-23**, 407–412.
24. Silberberg, T. (1980). Shape discrimination using Walsh descriptors. M.Sci. thesis, Ben-Gurion Univ. of the Negev, Israel.
25. Ahmed, N., and Rao, K. R. (1975). "Orthogonal Transforms for Digital Signal Processing." Springer-Verlag, Berlin.
26. Thomas, D. W. (1973). Burst detection using the Haar spectrum. *Proc. Theory Applic. Walsh Functions, Hatfield Polytechnic, England.*
27. Ahmed, N., Natarajan, T., and Rao, L. R. (1973). Some considerations of the Haar and modified Walsh–Hadamard transform. *Proc. Symp. Applic. Walsh Functions, Washington, D.C.*, AD763000, pp. 91–95.
28. Ahmed, N., and Rao, K. R. (1970). Spectral analysis of linear digital systems using BIFORE. *Electron. Lett.* **6**, 43–44.
29. Pratt, W. K., Krane, J., and Andrews, H. C. (1969). Hadamard transform image coding. *Proc. IEEE* **57**, 58–68.
30. Crowther, W. R., and Rader, C. M. (1966). Efficient coding of Vocoder channel signals using linear transformation. *Proc. IEEE* **54**, 594–595.
31. Campanella, S. J., and Robinson, G. S. (1970). Analog sequency composition of voice signals. *Proc. Symp. Applic. Walsh Functions, Washington, D.C.*, AD707431, pp. 230–237.
32. Harmuth, H. F. (1968). Sequency filters based on Walsh functions. *IEEE Trans. Electromag. Compat.* **EMC-10** 293–295.

33. Harmuth, H. F. (1970). Survey of analog sequency filters based on Walsh functions, *Proc. Symp. Applic. Walsh Functions, Washington, D.C.,* AD707431, pp. 208–219.
34. Golden, J. P., and James, S. N. (1971). LCS resonant filters for Walsh functions. *Proc. IEEE Fall Electron. Conf., Chicago,* pp. 386–390.
35. Nagle, H. T. (1972). Resonant sequency filters in the Z-doman. *Proc. Symp. Applic. Walsh Functions, Washington, D.C.,* AD744650, pp. 193–197.
36. Vandivere, E. F. (1970). A flexible Walsh filter design. *Proc. Symp. Applic. Walsh Functions, Washington, D.C.,* AD707431, pp. 3–6.
37. Beauchamp, K. G., and Yuen, C. K. (1973). "Digital Methods for Signal Analysis." Geo. Allen & Unwin, London.
38. Davenport, W. B., and Root, W. L. (1968). "Random Signals and Noise." McGraw-Hill, New York.
39. Pratt, W. K. (1971). Generalised Wiener filtering computation techniques. *IEEE Trans. Comput.* **C-21,** 636–641.
40. Rumatowski, K., and Sawicki, J. (1975). Using the discrete Walsh and Haar transforms in digital filtering. *Symp. Theory Applic. Walsh Functions, Hatfield Polytechnic, England.*
41. Rao, K. R., Narasimhan, M. A., and Revuluri, K. (1975). Image processing by Hadamard/Haar transform. *IEEE Trans. Comput.* **C-24,** 888–896.
42. Eghbali, H. J. (1980). Image enhancement using a high sequency ordered Hadamard transform filtering. *Computer Graphics* **5,** 23–29.
43. Gubbins, D., Scollar, I., and Wisskirchen, P. (1971). Two dimensional digital filtering with Haar and Walsh transforms. *Annal. Geophys.* **27,** 85–104.
44. Pratt, W. K. (1971). Linear and non-linear filtering in the Walsh domain. *Proc. Symp. Applic. Walsh Functions, Washington, D.C.,* AD727000, pp. 38–42.
45. Brown, C.G. (1970). Signal processing techniques using Walsh functions. *Proc. Symp. Applic. Walsh Functions, Washington, D.C.,* AD744650, pp. 138–146.
46. Kahveci, A. E., and Hall, E. L. (1972). Frequency domain design of sequency filters. *Proc. Symp. Applic. Walsh Functions, Washington, D.C., 1972,* AD744650, pp. 198–209.
47. Thompson, K. R., and Rao, K.R. (1977). Analysing a biorthogonal information channel by the Walsh–Hadamard transform. *Comp. Electr. Eng.* **4,** 119–132.
48. Karpola, F., and Jernigan, M. E. (1980). Extraction of periodic signals using the Haar transform. *IEEE Int. Conf. Cybernet. Soc.,* pp. 318–322.
49. van Cleave, J. (1977). Adaptive pulse processing means and methods. U.S. patent 4,038,539.
50. van Cleave, J. (1980). A Walsh pre-processor. NTIS Report AD-A091 188/3, Washington, D.C.
51. Rumatowski, K. (1983). Some aspects of the two-dimensional nonrecursive Walsh filtering. *Conf. Neuvieme Colloque sur le Traitement du Signal et ses Applications, Nice, France,* pp. 619–622.
52. Pratt, W. K. (1972). Walsh functions in image processing and two-dimensional filtering. *Proc. Symp. Applic. Walsh Functions, Washington, D.C.,* AD744650, pp. 14–22.
53. Thomas, D. W., and Wilkins, B. R. (1970). Determination of engine firing rate from the acoustic waveform. *Electron Lett.* **6,** 93–96.
54. Rao, P. R. (1980). Walsh domain filtering of finite discrete two-dimensional data. *IEEE Symp. Electromag. Compat., Baltimore, Maryland,* pp. 294–296.
55. Cooper, G. R., and McGillem, C. D. (1967). "Methods of Signal and System Analysis." Holt, Rinehart, New York.
56. Kitai, R. (1975). Synthesis of periodic sinusoids from Walsh waves. *IEEE Trans. Instrum. Meas.* **IM-24,** 313–317.

57. Chen, B. D., and Sun, Y. Y. (1980). Waveform synthesis via inverse Walsh transform. *Int. J. Electron.* **48,** 243–256.
58. Brown, W. O., and Elliott, A.R. (1972). A digital instrument for the inverse Walsh transform. *Proc. Symp. Applic. Walsh Functions, Washington, D.C.,* AD744650, pp. 68–72.
59. Beauchamp, K. G. (1973). Waveform synthesis using Fourier and Walsh series. *Proc. Theory Applic. Walsh Functions, Hatfield Polytechnic, England.*
60. Alexits, G. (1961). "Convergence Problems of Orthogonal Series." Pergamon, New York.
61. Sivak, G. (1979). The Haar transform: Its theory and computer implementation. U.S. Army Armament Research and Development Command Paper AD-AO70518/6. NTIS, Washington, D.C.

Chapter 4

Hardware Techniques

4.1 Introduction

Preceding chapters have considered the mathematical characteristics of Walsh and related sequency series and their generation and transformation through computer software. Many real-time applications, particularly those concerned with communications or on-line signal processing, require special-purpose hardware either to achieve the required operating speed or to achieve system efficiency. The availability and cheapness of general-purpose microprocessors and microelectronic designs further enhance the attractiveness of achieving generation and transformation of sequency series in this way [1]. In this chapter a number of the design and performance features of these developments will be considered. Applications of such designs will form the topics of subsequent chapters.

The basis of the use and transformation of the sequency functions lies in methods of series generation, and this subject will be considered first. One application, considered here for convenience, is the use of hardware series generation in waveform synthesis. Digital synthesis of commonly-used waveforms for laboratory testing (sinusoid, square-wave, ramp and triangular waveforms) has a number of advantages over continuous generation, and the use of the Walsh series for this purpose confers an economy in constituent logic and operation over a wide frequency range. Other requirements for sequency generators are in transmission by Walsh carriers, in real-time spectral measurement, sequency multiplexing and coding and in image processing.

For certain computational purposes, such as real-time digital filtering, image transmission and pattern recognition there is considerable interest in hardware transformation which can outclass computer software transformation in terms of speed and parallel operation. The advantages in simplicity provided with the sequency algorithms described earlier are carried forward into hardware design, and in describing these their incorporation into *single-chip* LSI logic will also be considered.

Finally many of the applications for sequency transformation require *analog-in, analog-out* operation, and so there is a need for a compact digital transformation device to replace the earlier analog and hybrid systems [2, 3]. The flexible 16 bit microprocessor with its ancillary analog–digital (A–D) and digital–analog (D–A) chips may be configured to meet this requirement, as can the newer charge-coupled devices now becoming available.

4.2 Walsh function generators

The sequency generators having the widest applicability are those generating a set of Walsh series, although in some cases the series are obtained by first generating a series of Rademacher functions. A generator that produces a set of N Walsh functions WAL(n, t), where $n = 0, 1, \ldots, N - 1$, is called an array generator. Ideally the generated waves will be orthogonal to each other, and some designs are better in achieving this than are others.

Two classifications of array generators are considered. The first generate fixed sets of Walsh functions WAL(n, t), where only the sequency range of the entire array is controlled externally. These *array generators* find use in multiplexing and signal processing. The second classification includes generators for which the sequency order n and/or the time interval t are controlled externally. These are known as *programmable generators*. Further sub-classifications of programmable generators can be defined, namely, serial programmable generators in which the time interval is fixed and the sequency order controlled and parallel programmable generators in which the sequency range is fixed and the time interval controlled.

4.2.1 Array generators

The simplest and most widely used method of generating a set of Walsh functions under sequency control is to derive these from products to Rademacher functions (Subsection 1.3.4). To generate 2^p Walsh functions p flip-flops may be connected to form a binary counter, their outputs providing a subset of p Walsh functions which are also Rademacher functions. The

4.2 Walsh Function Generators

remaining $(2^p - p)$ Walsh functions are then obtained from logical operations on the Rademacher functions.

An example of Harmuth's array generator [4] is shown in Fig. 4.1. The binary counters C_1 to C_4 produce the four Rademacher functions WAL(1, t), WAL(3, t), WAL(7, t) commencing with RAD(3, t) = WAL(7, t). From these functions, by using suitable multiplicative exclusive-OR gates, may be generated a complete set of the first 16 Walsh functions. The logical values of the functions generated are 0 and 1, corresponding to the $+1$ and -1 required by the Walsh function definition.

A principal requirement of a Walsh array generator is that the output functions exhibit a high degree of orthogonality. Orthogonality error may be caused by (1) propagation delays in the integrated circuits, (2) tolerances of

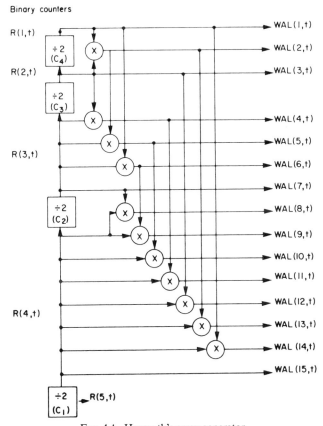

FIG. 4.1. Harmuth's array generator.

these delay times for the individual circuits and (3) differences between the rise and fall times of the integrated circuit outputs. Thus variable time delays can occur owing to transmission of the counteroutputs through different numbers of gates of non-identical performance. In addition, and as a consequence of these delays, spurious pulses can arise from modulo-2 addition of two unsynchronised signals.

A considerable improvement is obtained if the individual functions derive from the outputs of separate flip-flops that are fed from a common clock. The first generator of this type was that of Boesswetter [5]. This generator minimises delay differences by using synchronous clocking so that orthogonality errors are due only to differences in flip-flop switching times. Boesswetter's principle has been extended by Besslich [6], who demonstrates the suitability of ECL flip-flops in array generators. In his design a constant gate delay is used irrespective of the number of stages used. This is in contrast with Harmuth's original design, which includes an additional delay for every doubling of range for the generated series.

Design requirements other than low orthogonality error are the maximum clock rate desired and the level of hardware complexity in the number of gating elements used. Generators designed from JK flip-flops may be constructed from two-level NAND gates or, with fewer logic modules, from cascaded exclusive-OR gates [7]. These designs are economical in construction but slow in operation due principally to sequential circuit delays [8]. For arrays based on ECL flip-flops many complementary pairs can be used and logical requirements reduced. The clocking rate can be higher, and in Fernandez's design based on the Besslich generator (Fig. 4.2) and using Schottky transistor–transistor logic (TTL) a clock rate of 50M Zps with an orthogonality error of 2.8 ns is obtained [9]. Unlike JK flip-flop circuits, those based on ECL flip-flops are prone to induced noise since the inputs of the internal logic gates connect directly to output terminals and, further, the design has no *self-healing* property. This is, at any time after an external reset, a momentary disturbance can give rise to random flip-flop output states which remain until the next reset. Kitai [8] suggests using a circuit rearrangement to avoid this; it increases slightly the logic complement.

In all of the above, generation of the Walsh functions has proceeded from the multiplicative property of the function, viz.,

$$\text{WAL}(n, t)\, \text{WAL}(m, t) = \text{WAL}(n \oplus m, t) \tag{4.1}$$

where, in practice, pairs of Rademacher functions are chosen for multiplication. An alternative design, based on a combinational network associated with a unit pulse generator, allows AND logic gates to replace exclusive-OR gates, with consequent freedom from induced noise problems. The principle of *added-pulse generation* is very simple and is shown in Fig. 4.3. Added-

4.2 Walsh Function Generators

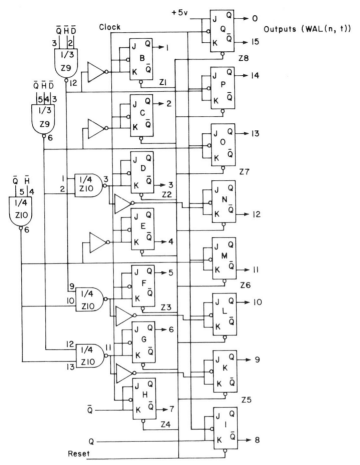

FIG. 4.2. Besslich array generator.

pulse generation employs a register-type counter in which a unit pulse is progressed through the counter at clock rate. In the case of the $N = 8$ function generator shown in Fig. 4.3, appropriate sets of $N/2$ pulses are selected and added in a four-input AND gate, one for each of the eight output functions. Comparison of the position of the selected pulses with the $+1$ level shown in the sequency-ordered set of Walsh function waveforms given in Fig. 1.4 will indicate how the Walsh function is built up from a combination of unit pulses. Sannino [10] has pointed out that this design can be very fast, limited only by delay times which characterise the logic family used. In his design a 16-WAL function generator is realised by using a square diode

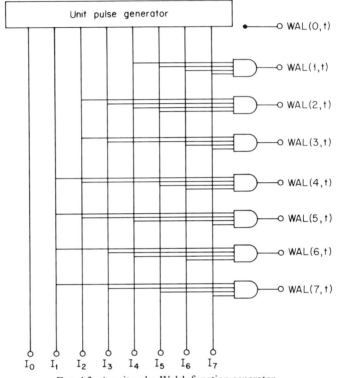

FIG. 4.3. A unit-pulse Walsh function generator.

matrix as an AND combinatorial network and TTL integrated logic modules for the register and other purposes. This method is inherently orthogonal, and in Sanino's implementation the error is shown as less than 10 ns.

4.2.2 Programmable generators

It is possible to implement the Walsh function definition given by Eq. (1.50) to produce the fully programmable generator. An example of this is given in Fig. 4.4. Here the r pairs of i and k parameters are input to $r - 1$ two-input exclusive-OR gates and r AND gates together with a r-input exclusive-OR half-adder required to carry out the modulo-2 addition and to form the output Walsh functions which are produced serially. In practise, however, little use is found for the fully-programmable generator, and attention has been directed to the serial and parallel implementations. Figure 4.4 can be rearranged to form a serial generator by simply replacing the k parameter

4.2 Walsh Function Generators

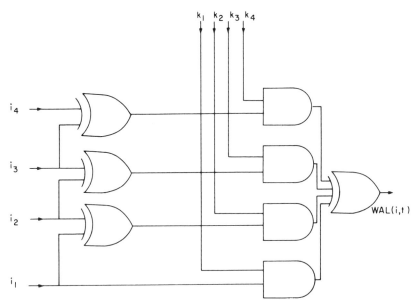

FIG. 4.4. A fully-programmable Walsh function generator.

inputs by a r-bit binary counter. A description of such a generator is given by Durrani and Nightingale [11].

One application of the serial programmable generator lies in the design of a sequential type of spectral analyser in which pairs of orthogonal SAL and CAL functions of a common sequency are needed [12]. It is essential in such a generator that the orthogonality error is reduced to a small value to avoid cumulative errors in the synthesis of generated functions forming the analyser waveform. This is achieved in a design due to Gaubatz and Kitai [13] by synchronous clocking of the ECL flip-flops producing the Walsh function outputs and the coding switches which are activated by the program input. Any orthogonality error due solely to differences in the switching times of the output flip-flops is then reduced to a few nanoseconds by the use of Schottky devices. Coding for the CAL and SAL functions is implemented through the use of a multiplier which combines the function of a synchronous 6-bit binary counter with selective decoding logic in a single integrated circuit. The logic circuits are arranged to gate the combination of each counter stage to form the correct CAL or SAL output sequence in accordance with a 6-bit input code, thus fulfilling the function of the unprogrammed Besslich generator shown in Fig. 4.2 but at a specified function level. Determination of those Rademacher or Walsh functions that must be multiplied together to form a specified Walsh function is considered by Davies [14] and Yuen [15].

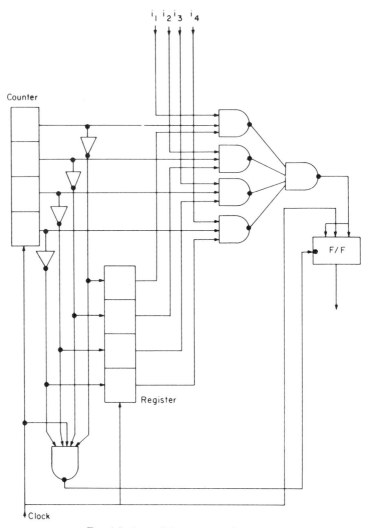

FIG. 4.5. A parallel programmed generator.

A number of parallel programmable generators have been developed for which the order i is externally controlled. Examples are the generators of Lee [16] and Peterson [17] which mechanise Eq. (1.51) by the use of a series of linear shift registers without reference to preceding members of the set. Peterson's method, like that of Swick [18], is based on symmetry considerations and is better suited to hardware construction since only a single one-directional shift register is needed.

Where N is large, the most effective designs, from the point of view of cost-effectiveness, are those in which Eq. (1.53) is implemented by using a Gray code counter. Yuen [15] has shown that a string of pulses representing sign changes of a Walsh function can be generated by combining the p output bits of a Gray code counter using an OR gate. The required Walsh function can then be reconstructed by triggering a bi-stable circuit from this string of pulses. A further advantage of the method is that the generator is unlikely to produce spurious outputs since only one non-zero pulse is produced by the p-bit Gray code incrementation. Thus there is little chance of two slightly unsynchronised pulses being combined to produce spurious outputs.

Generation of Gray code increments can be made by using a binary counter, as shown by Lebert [19], so that a complete parallel generator of this type may be realised by using standard integrated circuit components. A design for $p = 4$ due to Yuen [20] is shown in Fig. 4.5. Yuen's design has a symmetric structure and is, therefore, capable of expansion to form a larger series. Due also to the parallel operation of the generator, it is capable of high-speed operation since the outputs of the counter and register pass only two gates before arriving at another clocked component.

It is sometimes useful to be able to generate any of the various ordering patterns of a Walsh series. A global generator producing sequency-, dyadic- or natural-order output through logical combinations of a basic Rademacher function generator is described by Tzafestas [21]. This combines the generator of the sequency functions by modulo-2 addition of Rademacher functions with the appropriate Gray code conversion and other combinatorial logic to give the selected ordering as either a serial or parallel output.

4.3 Transformation

It will have become apparent from our study of fast Walsh signal flow diagrams in Chapter 2 that software implementation is simplified if the only transformation procedure used is the normal two-node butterfly (Fig. 2.4a). This is also true for hardware implementation, but here, since we are likely to be concerned with speed, the sequential routing of pairs of data values through a single hardware section would be very slow. If we consider τ as the time to complete one butterfly operation, then, in using a C-T fast transformation, $\tau n 2^{n-1}$ s would be needed for a 2^n point transformation (neglecting intermediate-stage storage operations). A parallel approach will be faster but will require n butterflies per stage, and each of the stage interconnections will be different. If a constant-geometry algorithm is used, then a single-stage hardware can be employed and 2^n sets of data routed n times through the one stage. Transformation will be achieved in approximately τn s. The fastest arrangement is a pipeline system whereby all n stages, each processing 2^n sets of data, are operated simultaneously. After the first n transformations a data transformation rate of one complete transformation for each τ, i.e., the time to complete one butterfly, will be achieved. This requires the maximum amount of hardware, namely, $n2^n$ butterflies. For some sequency function series such as the Haar function, the fast algorithm can employ fewer butterflies and hence faster pipelining. This is not an advantage with parallel processing, however, since this requires the maximum number of butterflies needed at any one stage.

In the following all three approaches will be considered in terms of hardware implementation. Included here also are microprocessor methods of transformation in which, although the transformation algorithm may be carried out serially, the availability of supporting software and interface chips makes for its rapid development as an on-line processing tool where fast throughput is not required. Designs for the fastest processing speed will employ parallel or pipeline methods and integrated circuits in their realisation. Recent developments are in the areas of programmable logic arrays which may be designed by efficient computer-aided design techniques and the charge-coupled devices which, because of their analog–digital nature, are particularly suitable for speech and image processing applications.

4.3.1 Parallel hardware systems

The fast Walsh transform was described earlier in matrix form as the product of the N coefficient data input x_i and an $N \times N$ coefficient Hadamard matrix H_n (Eq. (2.27)); H_n can be factored in three different ways. One of these leads to the familiar C–T algorithm which forms the basis of many software transformation programs (e.g., Figs. 2.6 and 2.7). The other two result in the product of n identical matrices and are therefore suitable for constant-geometry parallel hardware systems. If we take, for example, the case where $N = 8$ we can express these factorisations as

$$H_8 = PPP = QQQ \qquad (4.2)$$

where

$$P = \begin{bmatrix} 1 & 1 & 0 & 0 & 0 & 0 & 0 & 0 \\ 0 & 0 & 1 & 1 & 0 & 0 & 0 & 0 \\ 0 & 0 & 0 & 0 & 1 & 1 & 0 & 0 \\ 0 & 0 & 0 & 0 & 0 & 0 & 1 & 1 \\ 1 & -1 & 0 & 0 & 0 & 0 & 0 & 0 \\ 0 & 0 & 1 & -1 & 0 & 0 & 0 & 0 \\ 0 & 0 & 0 & 0 & 1 & -1 & 0 & 0 \\ 0 & 0 & 0 & 0 & 0 & 0 & 1 & -1 \end{bmatrix} \qquad (4.3)$$

and

$$Q = \begin{bmatrix} 1 & 0 & 0 & 0 & 1 & 0 & 0 & 0 \\ 1 & 0 & 0 & 0 & -1 & 0 & 0 & 0 \\ 0 & 1 & 0 & 0 & 0 & 1 & 0 & 0 \\ 0 & 1 & 0 & 0 & 0 & -1 & 0 & 0 \\ 0 & 0 & 1 & 0 & 0 & 0 & 1 & 0 \\ 0 & 0 & 1 & 0 & 0 & 0 & -1 & 0 \\ 0 & 0 & 0 & 1 & 0 & 0 & 0 & 1 \\ 0 & 0 & 0 & 1 & 0 & 0 & 0 & -1 \end{bmatrix} \quad (4.4)$$

The first of these **PPP** results in a constant-geometry flow diagram in which adjacent pairs of data elements are chosen for addition or subtraction and stored in consecutive locations for subsequent processing (Fig. 2.14). The second **QQQ** involves fetching pairs of data elements separated by $N/2$ locations and, after adding or subtracting, storing the results in adjacent locations for subsequent processing (as in stages A and B of Fig. 2.13).

In either case, since the interconnection pattern is the same for all stages of the computation, the hardware can be reduced to one stage by feeding the output back to the input and repeating the operation $\log_2 N$ times. This is shown in Fig. 4.6, in which the feedback path indicates that the output of each arithmetic element is fed back to the corresponding point above the element. In the case of $N = 8$ shown, new data would be entered through the multiplexer (MUX) at every third operation and the transformed data would be available in the output register at this time.

An early implementation of the first form of the constant-geometry sys-

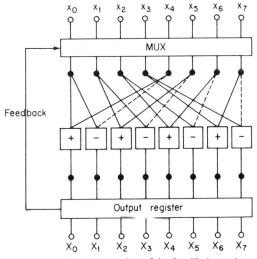

FIG. 4.6. An iterative implementation of the fast Hadamard transform.

tem **PPP** was carried out by Elliott and Shum [22], who used discrete logic. This employs a data-ordering register to drive a series of add/subtract logic units and carries out a recursion of the results from the output of the addition units to the inputs of the register. The procedure requires p transfers of data around the recursive loop to complete the Walsh transformation in natural order. Rearrangement to obtain sequency order requires the addition of a new set of registers at the output of the addition units with appropriate cross connections to reorder the data.

Another constant-geometry implementation, due to Geadah and Corinthios [23], also uses discrete logic to obtain a Walsh transformation. In their design the input memory holding the first N coefficient values to be transformed consists of a set of long dynamic shift registers which are divided into two halves, each $N/2$ bits long. Pairs of data values, separated by $N/2$ bits, are added or subtracted before being entered into a similar set of shift registers forming the output memory. The words in the input memory are then shifted 1 bit to the right and the procedure repeated for every successive data pair held in the registers x_i, $x_{i+N/2}$ until $x_{N/2-1}$ and x_{N-1} are reached. The contents of the output memory are then fed back to the input memory, one stage of the constant-geometry algorithm having been completed. Subsequent iterations are carried out, and, after p iterations are completed, the output coefficients can be read directly from the output memory. This gives the output in dyadic order. Some interstage shuffling of data values is required for natural or sequential order (see Fig. 2.15).

A different shift-register approach is described by Muniappan and Kitai [24]. Here the incoming data are stored in a permuted sequence in an auxiliary memory and the data sent periodically to the transformation processor. The processor is based on the **QQQ** form of the algorithm and is illustrated in Fig. 4.7. It consists of four shift registers SR1–4, each $N/2$ bits

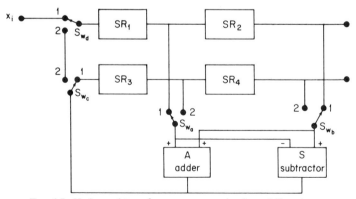

FIG. 4.7. Hadamard-transform processor using long shift registers.

long, an adder A, a subtractor S and four switches Sw_a, Sw_b, Sw_c and Sw_d. Initially the switches are in position 1. Input data x_i, $i = 0, 1, \ldots, N-1$, is fed into SR_1–SR_2, considered here as a single N-bit register. At the completion of input-data acquisition, Sw_d changes to position 2 and the first iteration commences. The data stored in SR_1–SR_2 are indexed from right to left so that the initial values stored, namely, x_0 in SR_1 and SR_2 are input as operands to A and S. The sum and difference outputs of A and S, respectively, are transferred to SR_3 through Sw_c. When this transfer is completed new data for A and S are fetched simultaneously by shifting the contents of SR_1–SR_2. This step is repeated $N/2$ times to complete the first half of the algorithm operation. Switches Sw_a, Sw_b and Sw_c then change to position 2 for the remaining half, with the roles of the upper and lower registers interchanged. The data inputs to A and S are fetched from SR_3–SR_4 and the outputs of A and S shifted into SR_1.

The entire process is repeated p times when the contents of the appropriate shift registers (upper pair for p even and lower pair for p odd) are available for output of the Walsh transformation in natural ordering.

The processing time is dependent mainly on the shifting rate of the registers. This is given by pN/f_s, where f_s is the shifting frequency. For $N = 1024$ real-time processing at an input rate of 1 MHz can be realised [24], which is somewhat faster than the Geadah and Corinthios [23] implementation. The input rate can be increased by a factor of p, with a pipeline organisation, by arranging p repetitions of the basic circuit.

Alternative outputs in natural, dyadic or sequency order are obtained by preceding the transform processor by a permutation module based on Berauer's [25] algorithm. It involves a permutation of the incoming data and performing a Hadamard (natural) transform of the permuted data sequence. This differs from the arrangement used by Geadah and Corinthios [23], which requires a reordering of the intermediate results, which vary with each iteration (see Fig. 2.15).

Several other parallel transformation arrangements have been described in which the recursive nature of the constant geometry is exploited. A hybrid analog–digital system has been developed by Carl and Swartwood [3] which effectively employs Eq. (4.3). However, in this case the single matrix is replaced by a matrix product $\mathbf{P} = \mathbf{S} \cdot \mathbf{B}$, where $\mathbf{B}$ represents a series of normal butterflies and $\mathbf{S}$ a simple interconnection algorithm. This enables the number of interconnections between circuits to be considerably reduced. The butterflies are realised by pairs of sample-and-hold amplifiers and sum/difference amplifiers (Fig. 4.8). They are interconnected through a gated set of sample-and-hold amplifiers serving a function similar to the interconnecting switches shown in Fig. 4.7.

An inspection of the flow diagrams associated with Eqs. (4.3) and (4.4)

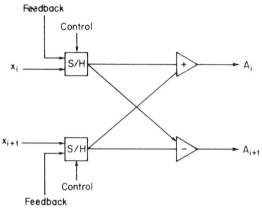

FIG. 4.8. A hybrid butterfly circuit.

shows that, if the combining unit at the stage output (B in Fig. 2.13) is replaced by an adder/subtractor unit, then the interconnections needed are halved. This was first suggested by Stone [26] in connection with parallel processing for the FFT. He called the resulting interconnection arrangement the *perfect shuffle* by analogy with the shuffling of two packs of m cards $(a, b, c, \ldots, m)$ and $(a', b', c', \ldots, m')$ into $(a, a', b, b', c, c', \ldots, m, m')$.

This constant-geometry transformation is shown in Fig. 4.9 for a Walsh transformation where $N = 8$. We can also define this interconnection as the mapping of the indices on the left of the diagram onto those on the right according to a permutation P such that

$$P(i) = 2i, \qquad 0 \leq i \leq N/2 - 1$$
$$P(i) = 2i + 1 - N, \qquad N/2 \leq i \leq N - 1 \qquad (4.5)$$

A useful feature of the perfect shuffle is that, if the indices are represented in binary form, then the ith element may be shuffled to its new position i' by cyclically rotating the bits one position to the left. Thus for

$$i = i_{n-1}2^{n-1} + i_{n-2}2^{n-2} + \cdots + i_1 2 + i_0$$
$$i' = i_{n-2}2^{n-1} + i_{n-3}2^{n-2} + \cdots + i_0 2 + i_{n-1} \qquad (4.6)$$

which is seen to agree with Eq. (4.5). This can simplify the hardware implementation of the required addressing routines.

Hardware design for the perfect shuffle implementation of Fig. 4.9 is based on an assembly of p identical single-stage units or a recursive procession through a single stage. An iterative processor has been described by Shirata and Nakatsuyama [27] which uses a combination of perfect shuffle networks and a set of adder/subtractor units to obtain fast Walsh transfor-

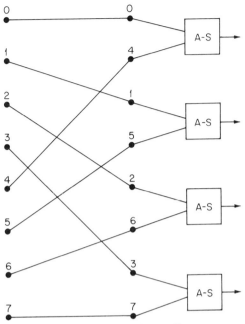

FIG. 4.9. The perfect shuffle.

mation in either natural, dyadic or sequency order. The shuffle networks in Shirata and Nakatsuyama's implementation are not, however, uniform, and 6 stages are required for $N = 16$, together with a set of 8 adder/subtractor units. Control of the output from the latter units enables the selection of intermediate outputs at each stage in the iteration so that the output can be in the selected order. The method of control is somewhat similar to that of Geadah and Corinthios [23], but the speed of transformation is increased.

A simpler design, also based on the perfect shuffle, is described by Nakatsuyama and Nishizuka [28]; it also provides the selection of any one of the three transformations through the use of a set of selection circuits following the adder/subtractor units. Here a set of such units are used after each of the identical perfect shuffle networks. Thus for $N = 8$ there are 3 networks and a total of 12 adder/subtractor units. Selection of dyadic, natural or sequency order is obtained as referred to earlier by selection of suitable intermediate-stage outputs for combination in the next iterative stage. Using adder/subtractor unit simplifies the switching, as is seen in Fig. 4.10. Here only a single output term is obtained, PAL(2, t), HAD(1, t) and WAL(4, t), shown as (a), (b) and (c), respectively in the diagram, which are selected by suitable switching control at the output of each stage. In this processor a set of selection logic modules follow each set of adder/subtractor units. The mod-

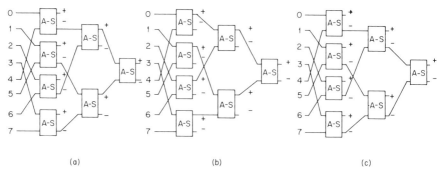

FIG. 4.10. Switching the perfect shuffle to obtain dyadic-, natural- and sequency-ordered output.

ules are controlled through a logic controller consisting of a binary-to-Gray-code converter and a bit-reversal network (see Subsection 1.3.5). The number of adder/subtractor units required is $pN/2$. Nakatsuyama and Nishizuka [28] also describe a pipeline processor in which this number is reduced by $N-1$ by the use of delay lines. Although the transformed word output is produced in parallel, which improves the speed of operation, only one component of the Walsh spectrum is calculated at a time.

4.3.2 Pipeline processors

Whilst the pipeline approach forms probably the fastest organisational solution to the problem of sequency transformation, it has not been widely accepted. This is due in part to the complex hardware needs, as discussed earlier, and due to the problems of control and signal-to-noise ratio in the transmission of data through the extended system logic. In order to reduce the complexity to a minimum all current developments for pipeline systems are highly modular.

A structure for the hardware realisation of the fast Walsh transform has been proposed by Ashouri and Constantinides [29]. The structure uses the basic delay line method employed by Grogisky and Wood [30] in their realisation of the FFT. A single stage in this system is shown in Fig. 4.11, which corresponds to a single butterfly operation (multiplication by a weighting function W^p has been omitted in this diagram and the process of multiplication of element values replaced by addition/subtraction to conform with the requirements of Walsh transformation). Here the signal input samples are presented sequentially to a pair of summing circuits arranged to add or subtract the current sample with a previous sample held in the delay register. Each register will need to hold $N-1$ sequential samples.

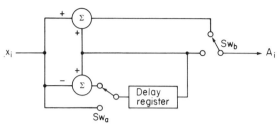

FIG. 4.11. A pipeline processor.

The transformation algorithm employed has the dyadic-ordered flow diagram shown in Fig. 2.6. Switching of the output connection to obtain the sum or difference of appropriate pairs of samples is made coincident with the switching of the delay register to the correct summer input. It will be apparent from Fig. 2.6 that different switching frequencies will be required for each of the p stages in the algorithm. The control mechanism for the switches is, in fact, a simple synchronised binary counter in which the binary number generated identifies the order of the sample currently being produced at the output. Thus the first transformed output sample will appear at the processor output as the Nth data sample is fed to the input of the processor, i.e., after all the registers are filled. Thereafter processing will occur at the rate of one completed transformed sample every clock period, which is determined solely by the operational time of one butterfly stage.

The structure and operation are somewhat similar to the pipeline implementation of the Muniappan and Kitai [24] processor, considered in the preceding section, but, since a constant-geometry algorithm is not employed, the individual modules are dissimilar. This dissimilarity will result in a higher chip count.

4.3.3 Microcomputer application

The current availability of low-cost microprocessors with substantive supporting software and interface chips has allowed the realisation of many sequency transformation and signal processing tasks on a dedicated computer basis. It is quite realistic to include the process of analog–digital and digital–analog conversion within the microcomputer organisation so that the digital equivalent of the earlier analog on-line processors can be obtained with better accuracy and much higher versatility.

Transformation using microprocessor systems generally relies on software serial operation, and consequently on-line systems are limited to sampling at a rate of only a few thousand samples per second, in contrast to the custom-designed parallel and pipeline systems described earlier. However, despite this considerably slower speed of operation, there are very many applications in which the microprocessor system matches the realised speed of data acquisition or in which the economy or flexibility of a microprocessor system is paramount. Examples are found in EEG processing, seismic signal analysis and various forms of process control.

To illustrate the possibilities and limitations of microprocessor application, two designs will be considered. The first is a low-speed system based on a popular and general-purpose microprocessor, the Intel 8080. This design is due to Durgan and Lai [31] and shows the advantages of implementation where it is based on the use of standard components and the selection of a suitable transformation algorithm. The second, also using an Intel microprocessor, employs some ancillary special-purpose logic to achieve a faster transformation enabling it to be employed in an on-line situation. In this particular design due to Muniappan and Kitai [32], the Walsh coefficients obtained from a band-limited signal are transformed into the corresponding Fourier coefficients, so that both spectra are available.

The first processor we consider is designed to carry out a Walsh transformation of a continuous analog signal where the frequency is limited to less than 50 Hz. An in-place algorithm is chosen for this application in order to reduce the memory storage requirements—an important consideration with microprocessor systems. The flow diagram was given in Fig. 2.9, where it is seen to be a modification of the C–T algorithm with the output obtained in bit-reversed sequency order. This is not a disadvantage where spectral averaging is required, since this can be carried out in bit-reversed order and the bit reversal carried out after the averaging has been completed.

A schematic diagram for the complete processor is given in Fig. 4.12. The input arrangements are important to ensure a reasonable digital representation of the continuous analog signal. Aliasing error is avoided by including a low-pass analog filter in the input channel and having a cut-off frequency determined by the highest frequency of interest contained within the incoming signal. Quantisation noise is minimised by choosing an A–D converter which has a wider dynamic range than is expressed by the sampling resolution. In this case a 12-bit converter is used for a 1024-point FWT. Transformation is carried out by using 3-byte integers. This eliminates possible integer overflow errors and the need to truncate the input data. Automatic scaling of intermediate-stage results maximises resolution and ensures minimum truncation error. This scaling is put into effect only when the maximum amplitude of the FWT output exceeds 23 bits or in the case of averaging outputs where the currently averaged spectra are greater than 23 bits.

A flow diagram for this design is shown in Fig. 4.13. Input parameters, such as sampling rate and number of points per transform, are set by the instrument control panel switches but could be entered as software parame-

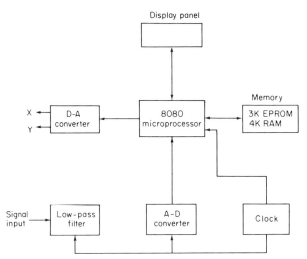

Fig. 4.12. A microprocessor Walsh transformer.

ters. The software includes provision for generating the power density spectrum from the Walsh spectrum by using the periodogram definition given by Eq. (3.15) and optional averaging of up to 99 sets of Walsh coefficients or power spectrum values. If this latter option is required, the input data may require truncation to ensure that numeric overflow errors will not occur. Checking of data overflow error is included and a suitable user diagnostic error provided.

Whilst the configuration of our second example also employs an Intel 8080, it is used in quite a different way in order to obtain the maximum transformation speed. The complete design is a self-contained spectral analyser including circuit arrangements to allow measurement over a very wide frequency range without range switching, but we shall only be concerned here with its transformation capability.

Choice of a suitable Walsh transformation algorithm is difficult where the principal requirements are speed, sequency-ordered output and simple implementation for a microprocessor having a limited memory. In Muniappan and Kitai's design the fast in-place algorithm described by Berauer [25] is used. This carries out a permutation of the input data and then carries out a fast transformation based on the FFT-derived algorithm shown in Fig. 2.7. This would normally provide an output in natural order, but with appropriate shuffling of the input data this can provide a sequency-ordered set of coefficients. The advantage is that only one kind of butterfly is involved, thus reducing program length and improving processing time through reduced flow control operations.

The permutation operation for a data length N is, as we would expect, a

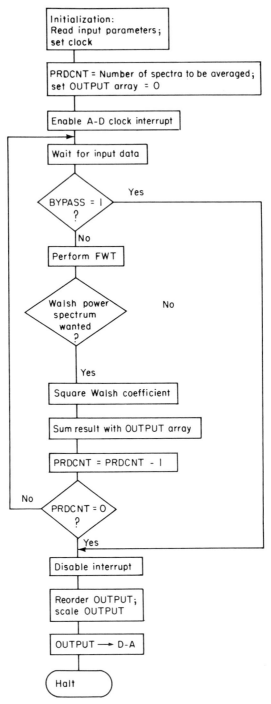

FIG. 4.13. Flow diagram for Fig. 4.12.

binary-to-Gray-code conversion followed by bit reversal

$$x_i(b) \rightarrow \overline{x_i(g)} \tag{4.7}$$

where $x_i(b) = x_0, x_1, \ldots, x_{N-1}$, is the input data sequence and $\overline{x_i(g)}$ the permuted data in the computer memory. The overbar indicates a bit reversal for the converted data. The relationship between $x_i(b)$ and $x_i(g)$ is as given previously in Eq. (1.42). In order to achieve a fast sequency-ordered transformation, it is necessary to carry out these operations by using hardware logic. Hardware logic uses a combination of direct memory access (DMA) for rapid access to the set of data required for a given transformation and a logic circuit to modify the normally sequential address from the DMA into the permuted address implied in Eq. (4.7). The configuration for the complete operation is shown in Fig. 4.14.

Transformation commences with program execution from the first memory address. The address register and terminal count register of the DMA controller are programmed with the starting address of the data and the number of samples to be acquired from the input signal. A DMA-request pulse, generated by the software, is applied and DMA takes control of the system bus to provide a set of 16-bit addresses in increasing order. These are permuted by the logic shown in the diagram, which acts to modify the six least significant bits and provide the correct shuffled order for the data operated on by the transformation algorithm. After the programmed number of samples have been processed by the DMA, the system bus is released

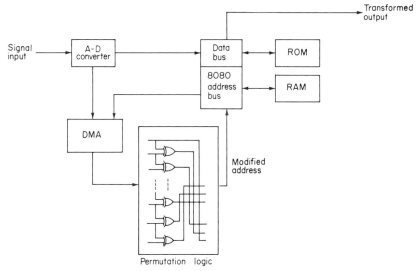

FIG. 4.14. A DMA-controlled processor.

and the processor carries out the FWT algorithm by using 16-bit arithmetic. Speed and accuracy of transformation are limited by the A–D converter and the word length of the processor. When an 8-bit A–D converter and an 8080 processor are used, the speed is stated as 4.5 kHz and dynamic range of 42 dB.

4.4 LSI application

The availability of integrated logic and LSI chips covering a wide variety of logic functions has led to their use as components in many individual designs for sequency generation, transformation and applications in which their cheapness and speed of operation confer many advantages, not the least being the simplification of design and assembly. To same extent this has meant that system design has developed most where LSI units have been readily available, such as in the use of the rate multiplier in Gaubatz and Kitai's [13] generator and DMA control in Muniappan and Kitai's [32] spectral analyser, whilst the use of RAM in data shuffling permits the use of complex addressing sequences required for alternative function ordering [24]. Many designs make use of RAM, ROM or EPROM for coefficient storage [32, 34], digital integrated circuits for fast add/subtract operations [24], registers and control logic to supplement or replace discrete logic circuits. There is a need for a *single-chip* Walsh function generator for use as an applications component, and design studies are being carried out on more complex LSI components for sequency applications.

In one of the studies carried out by Nick [34] of IBM, the design of a complete Walsh function spectral analyser is described. This, like the earlier Brown and Elliott [35] design makes use of an integral Walsh function generator derived from a gated Rademacher function generator by using the matrix relationship $x_i \cdot \mathbf{W}$, where $\mathbf{W}$ is an nth-order Walsh matrix. The design is a direct transformation and is not based on sparse matrices fast transformation.

The transformation of a binary input signal into its Walsh series and the transmission of a previous Walsh transformation are accomplished simultaneously within a given clock cycle. This clock cycle consists essentially of the number of sub-intervals which coincide with the desired resolution of the input signal and is thus proportional to the Walsh matrix order N.

The complete design includes registers storing the vector product representing Walsh function amplitudes and a set of line drivers for transmission of the series at a high data rate over coaxial lines. For a logic circuit family having a packaged delay of 15 ns per logic stage (AND–OR gates), the speed and complexity for a different size of the input series is given by Nick and reproduced in Table 4.1.

4.4 LSI Application

TABLE 4.1. Performance and complexity of a single-chip Walsh spectral analyser. (After Nick [34].)

	$N = 8$	$N = 16$	$N = 32$	$N = 64$
Cycle time (nanoseconds)	629	1387	3232	7429
Number of logic elements	1046	3040	9548	32019

Any specially-designed chip for sequency function generation or transformation (or for any signal processing function using other techniques) will have to satisfy a large demand if the present high costs of individual chip design are to be recovered. Hence there is considerable interest in programmable devices which have a wider application outside the applications discussed here. The application of microcomputers was discussed earlier and has the attraction of economy of design. The logic and other functions required in sequency work can be achieved by serially executing instructions stored in a memory module and logic changes made by simply altering the controlling program.

An alternative to the microprocessor for certain group logic operations has recently made its appearance and appears to be very suitable for sequency operations. This is the *programmable logic array* (PLA), which is a single chip device combining the properties of a logic memory with combinatorial logic and internal feedback to maintain the programmed states. Parallel operation on the incoming data makes it much faster than microcomputer operation, and changes to the internally programmed logic can be achieved rapidly by using standard CAD (computer-aided design) tools.

4.4.1 PLA devices

A schematic diagram for a programmable logic array is shown in Fig. 4.15. Essentially the array logic can be said to be embodied in a memory-like structure wherein the input data are used as an address and the output bits that are selected are combined to yield a word which is the value of a function stored in the array for a particular combination of input bits. Thus the selection of output values is carried out not directly, but through an associated or translation function. This results in a *read-only* structure which can be programmed to form both sequential and combinational logic—two essential elements in a transformation sequence [36].

The combinational logic is implemented by means of a sum-of-product function carried out by the AND and OR arrays operating is cascade. This is basically a look-up table structure in which the AND array forms the look-up

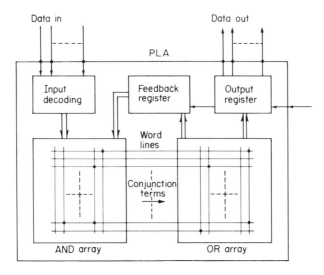

FIG. 4.15. A programmable logic array.

library and the OR array forms the resulting output for the operation. The AND array is programmed to search for a match between the logic values in the input register and the logical pattern made by the array interconnections. These can be 1, 0 or no connection. Selected output words are conveyed to the OR array, which combines the conjunctive terms found to form the required output. This process is indicated by the location of the dots in the skeleton arrays shown in the diagram.

Since the PLA has this associative property, the word lines search simultaneously for a match in the AND array and thus provide a parallel processing capability. This is in contrast with a ROM store, in which only one unique word line can become active at a time.

Sequential operation of this process of selection and combination is obtained through a feedback path, internal to the chip, from the OR to the AND array. The internal feedback register also provides a facility for implementing counters or changing the routine of sequential operations. This latter may be required, for example, where an alternative sequency ordering is desired. Then the read-array output lines can be fed back as search-array input lines via this register and hence implement combinatorial as well as sequential logic.

It is apparent from this that a PLA implementation contains the essential ingredients for successful sequency transformation, namely, logical selection of data pairs in accordance with a stored logic routine, parallel addition/subtraction, iteration through a feedback path and counting logic.

Design methodology, referred to as *personalization* of a PLA, consists of arranging the connecting paths in the AND and OR matrices during the final productivity stages of the package. This is achieved by mask programming as part of the chip production routine or by fusible diode links within the structure as found with a conventional PROM.

An implementation is described by Smith [37] in which CAD methods are used to program the PLA to carry out in-place Walsh transformation. A device simulation is first obtained and is followed by a software routine to translate the CAD model into a language which describes data used in mask generation for VLSI chip production. Full details of the design methodology are given by Smith [38].

Sequency methods are also proposed by Lechner [39] for the design of the PLA system itself. It will be apparent from the operation of the PLA as described earlier that the number of product terms is determined by the size of the input gate array. It is possible, however, to limit the number of required product terms, and hence input gate array size, by adopting a method of coding for the combinatorial logic which stores a sequency function of the Boolean logic expression. To do this Lechner computes the logical autocorrelation of a related Boolean expression through the WHT and then selects only the highest values of these for matrix inversion. The resulting matrix is used as an encoding matrix for the PLA. The method is shown to give a reduction of about 30% in input gate array size, although additional encoding logic is also required.

4.4.2 Charge-coupled devices

Charge-coupled devices (CCD) offer a number of very attractive advantages for signal processing due to their essential simplicity, cheapness and combined analog–digital operation [40]. The device is essentially a shift register formed on a silicon substrate as a string of closely spaced solid-stage capacitors. A CCD can store and transfer analog-charge signals that may be introduced electrically or optically. It can be considered as the equivalent to an analog-tapped delay line having digital control — a combination which results in a useful signal-processing tool combining the best features of digital and analog techniques.

In most applications the CCD operates as the transversal filter shown in Fig. 4.16. It consists of M delay stages D together with circuits to carry out the weighted summation of the node voltages V_k. The signal is non-destructively sampled at each stage and clocked at clock frequency f. The node voltages are multiplied by weighting coefficients h_k, $k = 0, 1, \ldots, M - 1$ and the products summed.

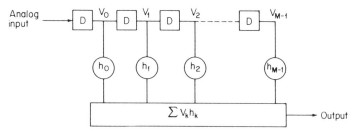

FIG. 4.16. A CCD transversal filter.

Very large numbers of delay stages are possible in a single device, and a 500-stage CCD is quite usual. This makes the device valuable for image processing and as a cost-effective method of achieving matched filtering. It is important to note that sampled analog signals are processed under digital control, which eliminates the requirements for analog–digital conversion in many cases. The limitations of a CCD lie in its dynamic range, linear charge transfer efficiency and leakage (which limits filter performance relative to the digital filter). These disadvantages are outweighed in many applications by CCD's low cost, particularly for large data processes such as image processing.

The CCD finds its main application in digital filtering, in spectral analysis and as a read-only memory [41]. As a transformation device for spectral analysis it is particularly effective in providing a significant potential saving over digital fast transformation. The way in which the CCD is used to implement a discrete transform is, however, quite different from the conventional hardware constructions considered earlier. In the digital fast transform algorithms we are interested in minimising the number of multiplications. This factor is no longer important with CCD since transversal filters can be constructed to carry out large numbers of multiplications simultaneously in real time. Instead we look for algorithms in which the major part of the computation is carried out by the transversal filter. Two well-known algorithms for this purpose are the chirp-Z transform and the prime transform [42]. These are, however, indirect means of achieving a transformation, and several direct means of obtaining a fast transform with CCD are described in the literature.

One of these, due to Yarlagadda and Hershey [43], carries out a fast Walsh transformation by using a version of the constant-geometry algorithm shown in Fig. 2.14. A simplified diagram of this is shown in Fig. 4.17 for $N = 8$. The diagram consists of the CCD register C, which receives its serial data input initially from a data loading sequence and subsequently via the feedback path shown. Output from the intermediate taps on the CCD is taken via a series of switches $S_1 - S_9$ to a summing stage which returns selected

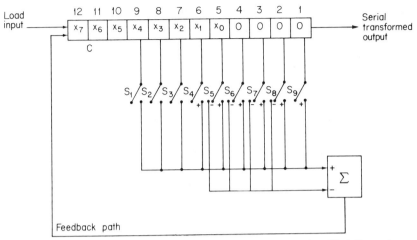

FIG. 4.17. Yarlagadda and Hershey's CCD Walsh transformer (simplified diagram).

pairs of summed outputs according to $x_k + x_{k+1}$ or $x_k - x_{k+1}$ to the data input of the CCD.

The transformation procedure commences by loading the N data samples into the first N spaces of the CCD register, which contains additional spaces to provide a working buffer ($N + 4$ in this diagram). Thereafter, data are entered via the feedback path shown and shifted along the register one position at each clock pulse. Thus after N clock pulses the data are in position as shown in the diagram with all the switches S_1–S_9 open. On the next clock pulse $N + 1$, the data are shifted along one position to the right and switches S_5 and S_6 closed to input the value of samples x_0 and x_1 into the addition stage and the value $x_0 + x_1$ entered via the feedback path into the vacant first CCD register position. Subsequent clock pulses load pairs of data into the CCD register until, after a total of $2N$ pulses, stage B of the flow diagram shown in Fig. 2.14 has been completed. The CCD register will now hold data as shown in Table 4.2. Note that some of the outputs from the CCD taps will need to be routed to the negative input of the summer to effect $x_0 - x_1$, etc. Thus the

TABLE 4.2. Data position in the CCD register after one iteration.

12	11	10	9	8	7
$x_6 - x_7$	$x_4 - x_5$	$x_2 - x_3$	$x_0 - x_1$	$x_6 + x_7$	$x_4 + x_5$
6	5	4	3	2	1
$x_2 + x_3$	$x_0 + x_1$	x_7	x_6	x_5	x_4

switches need to be controlled by a ternary level (0, +, −) controller. This is achieved by additional digital registers and control logic, neither of which is shown in the diagram.

Further stages of the transformation algorithm are completed in a similar manner. The contents of the CCD register shown in Table 4.2 are now in the correct position to repeat the procedure described and produce the output of the next stage C. After $3N + 4$ clock pulses the completed transformed Walsh output is available in serial natural order from the output of the CCD register. An alternative implementation of a CCD hardware transformation is also suggested by Yarlagadda, which requires less controlling logic but uses more CCDs operating at two different clocking speeds.

Light-activated CCDs are used in video sensor (camera) systems and transform techniques used to achieve reduced transmission bandwidth. Several methods have been proposed in which a two-dimensional Walsh transformation is employed. Merola *et al.* [44] describe a method for achieving this at the focal plane of the image which uses the minimum of supporting logic.

The solid-state CCD image sensor employed consists of a rectangular array of split-capacitance devices (Fig. 4.18), which we can regard as a storage well for the accumulated charge arising from the light source. This charge can be accessed through X and Y lines connected to the CCD elements in a manner similar to that used for memory devices. In Merola *et al.*'s application the charge is not accessed directly, however, but is first modified by Walsh coding to produce a line-by-line image output transformed into the Walsh domain, which is shown in Fig. 4.19. The Walsh generator is used to apply appropriate voltage charges to the column electrodes to obtain the indicated sums on the row lines. Thus the Walsh generator potentials E_1–E_4

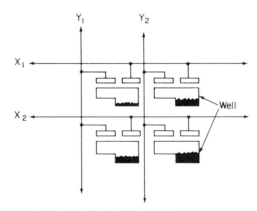

FIG. 4.18. A solid-stage CCD image sensor.

4.4 LSI Application

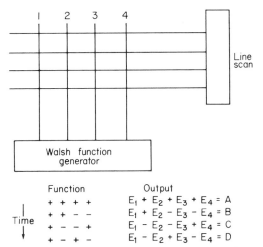

FIG. 4.19. Simplified diagram of a CCD Walsh transformer.

for WAL(0, t) are first applied to the columns such that the sum of pels $1 + 2 + 3 + 4$ gives the transformed image potential output corresponding to the first Walsh function of the series. (Here shown as a 4×4 matrix.) The column potentials are then modified to give a positive charge contribution from pels 1 and 2 and a negative charge from pels 3 and 4, so that the next row read-out will be $1 + 2 - 3 - 4$, corresponding to the second Walsh function WAL(1, t). The third and fourth transformed image potentials are generated in a similar manner.

In the application of this principle to the design of a CCD focal plan processing chip, the transformations in the horizontal direction are carried out as described above whilst the row summations required to complete the two-dimensional transformation (see Section 2.8) are performed by ancillary hardware. The processor is essentially a serial device and sequential read-out is arranged as a series of 32 sub-blocks of 2×2-pel image areas (the array contains 128 rows and 128 columns). Arranging read-out in this way permits some parallel processing of data from the sub-blocks. Referring again to the simplified diagram of Fig. 4.19, we see that, when the Walsh code generator sets up a particular coefficient for read-out, this becomes available at the output of each row. Thus several rows can be read simultaneously, allowing parallel processing to take place. Further in the process of taking an image transform, the signal charge at each pel is read out repeatedly, once for each transform value. This results in a signal-to-noise improvement of $N^{1/2}$ due to the averaging effect of repeated read-out.

A somewhat similar transformation method is described by Shaw and Westgate [45]. It uses an external ROM or microprocessor control to implement the Walsh transformation through generation of the required timing sequence. This also takes advantage of the simplification of dividing the CCD array into a number of sub-images so that a measure of parallel processing can be achieved. A multiplexing and demultiplexing scheme is used to select the desirable input sub-images to process and transmit.

The difficulty in handling subtraction of charges within the image sensor element, which was achieved in the preceding example through the use of a split capacitance device, is solved here in a different way. The Hadamard transformation matrix $\mathbf{H}$ is decomposed into

$$\mathbf{H} = \mathbf{A} - \overline{\mathbf{A}} \qquad (4.8)$$

where the components of $\overline{\mathbf{A}}$ are the complements of the corresponding components of $\mathbf{A}$ (Fig. 4.20). Hence we can write for the transformation of the matrix input $\mathbf{x}$

$$\mathbf{X} = \mathbf{H} \cdot \mathbf{x} \cdot \mathbf{H} = \mathbf{X}_+ - \mathbf{X}_- \qquad (4.9)$$

where

$$\mathbf{X}_+ = \mathbf{A} \cdot \mathbf{x} \cdot \mathbf{A} + \overline{\mathbf{A}} \cdot \mathbf{x} \cdot \overline{\mathbf{A}} \qquad (4.10)$$

and

$$\mathbf{X}_- = \mathbf{A} \cdot \mathbf{x} \cdot \overline{\mathbf{A}} + \overline{\mathbf{A}} \cdot \mathbf{x} \cdot \mathbf{A} \qquad (4.11)$$

The Walsh transform operation is decomposed into two parts, both of which need only the summation operation (easily achieved within the CCD storage well) with selected sets of pels for pel sub-image. The subtraction $\mathbf{X}_+ - \mathbf{X}_-$ is carried out externally to the CCD device. Since only ones and zeros are present in $\mathbf{A}$ and $\overline{\mathbf{A}}$, it is much easier to use an external ROM or microprocessor to generate the required sequence of control operations.

An even simpler method is used by Roberts *et al.* [46] in their implementation of a single-dimensional CCD transform device. Here the taps on the CCD are connected to a resistor network which provides N simultaneous transversal filters with responses determined by the conductances which weight the contributions made by the stored signal samples to a summing amplifier (Fig. 4.21). The circles in this diagram denote resistive intercon-

$$\mathbf{H} = \begin{bmatrix} 1 & 1 & 1 & 1 \\ 1 & -1 & 1 & -1 \\ 1 & 1 & -1 & -1 \\ 1 & -1 & -1 & 1 \end{bmatrix} = \overset{\mathbf{A}}{\begin{bmatrix} 1 & 1 & 1 & 1 \\ 1 & 0 & 1 & 0 \\ 1 & 1 & 0 & 0 \\ 1 & 0 & 0 & 1 \end{bmatrix}} - \overset{\overline{\mathbf{A}}}{\begin{bmatrix} 0 & 0 & 0 & 0 \\ 0 & 1 & 0 & 1 \\ 0 & 0 & 1 & 1 \\ 0 & 1 & 1 & 0 \end{bmatrix}}$$

FIG. 4.20. Decomposition of a Hadamard matrix $\mathbf{H}$ into $\mathbf{A}$ and $\overline{\mathbf{A}}$ matrices.

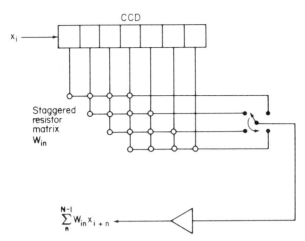

FIG. 4.21. Roberts resistor matrix CCD transformer.

nections in the non-contacting lattice of conductors. The switch commutes synchronously with the CCD clock, so that the scalar product of each row of W with the same block of data is output on successive clock cycles until the switch returns to its first position, at which point the processing of a new block of N data is commenced. The diagram omits the extra horizontal lines required to produce negative tap weights; these are summed separately and subtracted in a further operational amplifier. This method is only suitable for small values of N, since N^2 resistors are required and a large array size would be prohibitively expensive to implement.

4.4.3 SAW application

Surface acoustic wave (SAW) devices form yet another solid-state processing technique which is now well established for frequency filtering and transformation based on the ability of the device to store energy information on a slowly propagating acoustic wave. There is some similarity and overlapping of functions with the charge-coupled devices described in the preceding section. However, with bandwidths of several hundred megahertz and dynamic ranges extending towards 100 dB, passive SAW devices form a strongly competitive signal processing element. Surface acoustic wave devices are produced by the same photolithographic processing used for metalisation in IC manufacturing and are thus compact, reliable and economical. The simplest SAW device is the non-dispersive delay line shown in Fig. 4.22. The line consists essentially of two interdigital transducer arrays taking the form of interlaced *combs* of metallic fingers mounted on a piezoelectric

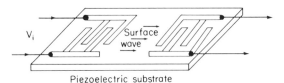

FIG. 4.22. A simple SAW delay line.

substrate, typically quartz or lithium niobate. An alternating voltage—the signal to be processed—is applied to one of these transducers, causing the material between the fingers to distort owing to the piezoelectric effect. This periodic strain leads to the propagation of an acoustic wave along the surface of the substrate to the second transducer. There the signal is detected by the inverse piezoelectric action of the transducer after a time delay L/v, where L is the separation of the transducers and v the velocity of the acoustic wave along the substrate. In many devices the role of the transducer is not only to convert from electric to acoustic energy, but also to filter the signal. This can be affected by adjusting the relative length or spacing of the fingers comprising the transducer. In Fig. 4.22 the surface wave is launched and detected without any intermediate interference. Various possibilities exist for redirecting and in some cases processing the signal in transit [47]. To achieve a multi-tap delay line, a periodic set of reflectors set normal to the incident beam or inclined at some other angle may be used. Alternatively, a set of interdigital transducers may be located at the tap positions.

This latter arrangement forms the basis of a domain transformation device which makes use of Hadamard matrix properties. It is of particular value in image processing applications for information compression, in which acoustic surface-wave-tapped delay lines are efficient for the achievement of such a transform in a low-cost real-time device.

The principle of a SAW matrix transform is illustrated in Fig. 4.23. A set of interdigital transducers are printed on a piezoelectric substrate in the form of a 4 × 4 matrix with one transducer for each position on the matrix and a

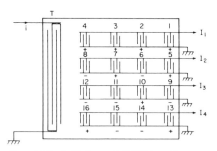

FIG. 4.23. A Hadamard SAW transformer.

further transducer T to launch the acoustic wave $i(t)$, where i represents the intensity of the image signal. The matrix operator H (Eq. (1.46)) is a square array of plus and minus ones and is implemented in this realisation by arranging to collect the output at each of the delay line taps with a phase of 0 or π rad. This is indicated in the diagram by the plus and minus signs at each transducer position. Thus a one-dimensional forward transform of the input signal $I = Hi$ is formed by line summation to give line output values $I_1 - I_4$. The first term I_1 is obtained by summing the electrical signal when the acoustic waveform launched by T is present under transducers 1, 2, 3, 4 at time t_0 to give $i_1 + i_2 + i_3 + i_4$. The second term I_2 gives $-i_5 + i_6 - i_7 + i_8$, etc., giving a parallel read-out transformed vector $I_1 - I_4$ at time t_0. Henaff [48] and Rebourg [49] describe practical applications of Hadamard SAW transform devices based on multiple operation of this basic transform. Fast serial output for each line may be obtained by shifting successive rows in the matrix by one position. A two-dimensional transformation is achieved by means of two applications of the transform; first along each row of the initial image (as in Fig. 4.23) and then along each column of the intermediate image. A Hadamard-like hardware transform has also been implemented by Tseng [50] who uses a surface wave device.

References

1. Yuen, C. K., Beauchamp, K. G., and Robinson, G. P. S. (1982). "Microprocessor Systems and Their Application to Signal Processing." Academic Press, London and New York.
2. Harmuth, H. F., Kamal, J., and Murty, S. (1974). Two dimensional spatial hardware filters for acoustic imaging. *Proc. Symp. Applic. Walsh Functions, Washington, D.C.*, pp. 94–125.
3. Carl, J. W., and Swartwood, R. V. (1973). A hybrid Walsh transform computer. *IEEE Trans. Comput.* **C22**, 669–672.
4. Harmuth, H. (1972). "Transmission of Information by Orthogonal Functions." Springer-Verlag, Berlin.
5. Boesswetter, C. (1970). Die Erzeugung von Walsh Funktionen. *Nachrichtentech Z.* **23**, 201–207.
6. Besslich, P. W. (1973). Walsh function generators for minimum orthogonality error. *IEEE Trans. Electromag. Compat.* **EMC-15**, 177–180.
7. Schreiber, H., and Sandy, G. F. (eds.) (1974). Applications of Walsh functions and sequency theory. *Proc. Symp. Applic. Walsh Functions, Washington, D.C.*, pp. 316–344.
8. Kitai, R. (1979). Walsh-function array generators: A comparison. *IEEE Trans. Electromag. Compat.* **EMC-21**, 153–154.
9. Fernandez, L. C., and Rao, K. R. (1977). Design of a synchronous Walsh function generator. *IEEE Trans. Electromag. Compat.* **EMC-19**, 407–410.
10. Sannino, M. (1978). Multiple-output Walsh function generator for minimum orthogonality error. *IEEE Trans. Instrum. Meas.* **IM-27**, 29–32.
11. Durrani, T. S., and Nightingale, J. M. (1971). Sequential generation of orthogonal functions. *Electron. Lett.* **7**, 385–387.

12. Siemens, K. H., and Kitai, R. (1969). Digital Walsh–Fourier analysis of periodic waveforms. *IEEE Trans. Instrum. Meas.* **IM-18**, 316–321.
13. Gaubatz, D. A., and Kitai, R. (1974). A programmable Walsh function generator for orthogonal sequency pairs. *IEEE Trans. Electromag. Compat.* **EMC-16**, 134–136.
14. Davies, A. C. (1972). On the definition and generation of Walsh functions. *IEEE Trans. Comput.* **C-21**, 187–189.
15. Yuen, C. K. (1973). Walsh function generation using Gray code. *Proc. Symp. Applic. Walsh Functions, Washington, D.C.*, AD763000, pp. 284–289.
16. Lee, J. S. (1970). Generation of Walsh functions as binary image groups. *Proc. Symp. Applic. Walsh Functions, Washington D.C.*, AD707431, pp. 7–11.
17. Peterson, H. L. (1970). Generation of Walsh functions. *Proc. Symp. Applic. Walsh Functions, Washington, D.C.*, AD707431, pp. 55–57.
18. Swick, D. A. (1969). Walsh function generation. *IEEE Trans. Info. Th.* **IT-15**, 167.
19. Lebert, F. J. (1970). Walsh function generation for a million different functions. *Proc. Symp. Applic. Walsh Functions, Washington, D.C.*, AD707431, pp. 52–54.
20. Yuen, C. K. (1973). Some programmable high-speed Walsh function generators. *Proc. Theory Applic. Walsh Functions, Hatfield Polytechnic, England.*
21. Tzafestas, S. (1976). Global Walsh function generators. *Electron. Eng.* **48** (585), 45–50.
22. Elliott, A. R., and Shum, Y. Y. (1972). A parallel array hardware implementation of the fast Hadamard and Walsh transforms. *Proc. Symp. Applic. Walsh Functions, Washington, D.C.*, AD744650, pp. 181–183.
23. Geadah, Y. A., and Corinthios, M. J. G. (1977). Natural, dyadic and sequency order algorithms and processors for the Walsh–Hadamard transform. *IEEE Trans. Comput.* **C-26**, 435–442.
24. Muniappan, K., and Kitai, R. (1982). Walsh spectrum measurement in natural, dyadic and sequency ordering. *IEEE Trans. Electromag. Compat.* **EMC-24**, 46–49.
25. Berauer, G. (1972). Fast in-place computation of the discrete Walsh transform in sequency order. *Proc. Symp. Applic. Walsh Functions, Washington, D.C.*, AD744650, pp. 272–275.
26. Stone, H. S. (1971). Parallel processing with the perfect shuffle. *IEEE Trans. Comput.* **C-20**, 153–161.
27. Shirata, K., and Nakatsuyama, M. (1980). The fast Walsh–Hadamard transform and processors by using new permutation networks. *Trans. IECE Jpn.* **J63-D**, 319–325.
28. Nakatsuyama, M., and Nishizuka, N. (1981). The fast Walsh–Hadamard transform and processors by using delay lines. *Trans. IECE Jpn.* **E64**, 708–715.
29. Ashouri, M. R., and Constantinides, A. G. (1977). A pipe-line fast Walsh–Fourier transform. *IEEE Conf. ASSP, Hartford, Connecticut*, pp. 515–518.
30. Grogisky, H. L., and Wood, G. A. (1970). A pipe-line fast Fourier transform. *IEEE Trans. Comput.* **C-19**, 1015–1019.
31. Durgan, B. K., and Lai, D. C. (1980). A microprocessor implementation of the fast Walsh transform. *IEEE Proc. Applic. Mini Microcomput., Philadelphia*, pp. 395–399.
32. Muniappan, K., and Kitai, R. (1979). Microprocessor-based Walsh–Fourier spectral analyser. *IEEE Trans. Instrum. Meas.* **IM-28**, 295–299.
33. Ahmed, N., and Hein, D. (1978). On a real-time Walsh/Hadamard/cosine transform image processor. *IEEE Trans. Electromag. Compat.* **EMC-20**, 453–457.
34. Nick, H. (1980). Binary logic Walsh function generator. *IBM Tech. Disclosure Bull.* **22**, 4650–4651.
35. Brown, W. O., and Elliott, A. R. (1972). A digital instrument for the inverse Walsh transform. *Proc. Symp. Applic. Walsh Functions, Washington, D.C.*, AD744650, pp. 68–72.
36. Fleisher, H., and Maissel, L. I. (1975). An introduction to array logic. *IBM J. Res. Develop.* **19**, 98–109.

37. Smith, E. G. (1980). Computer-aided design of a PLA-implemented fast Walsh–Hadamard transform device. *IEEE Proc. Int. Conf. Pattern Recognition, 5th,* pp. 180–182.
38. Smith, E. G. (1980). CAD design of modularised fast Walsh–Hadamard transformation device. *IEEE Proc. Int. Conf. Circ. Comput.,* pp. 916–920.
39. Lechner, R. J., and Moezzi, A. (1983). Synthesis of encoded PLA's. *Int. Conf. Fault Detection Spectral Techniques, Boston,* pp. 1.3–1.11.
40. Buss, D. D., Tasch, A. F., and Benton, J. B. (1979). CCD Applications to Signal Processing, *in* "Charge-Coupled Devices and Systems" (M. J. Howes and D. V. Morgan, (eds.)), pp. 79–88. Wiley, New York.
41. Melen, R., and Buss, D. (eds.) (1977). "Charge-Coupled Devices: Technology and Applications." IEEE Press, New York.
42. Buss, D. D., Brodersen, R. W., Hewes, C. R., and Tasch, A. F. (1975). Comparison between the CCD C7T and the digital FFT. *Proc. Naval Electron Lab. Center Int. Conf. Applic. Charge-Coupled Dev.,* pp. 267–281.
43. Yarlagadda, R., and Hershey, J. E. (1981). Architecture of the fast Walsh–Hadamard and fast Fourier transforms with charge transfer devices. *Int. J. Electron.* **51,** 669–681.
44. Merola, P. A., Michon, G. J., Burke, H. K., and Vogelsong, T. L. (1977). Charge injection device focal plane processor for video bandwidth compression. *Proc. Soc. Photo-Opt. Instrum.* **119,** 115–120.
45. Shaw, V. M., and Westgate, C. R. (1980). A CCD image pre-processor for Hadamard transform operations. *Proc. IEEE* **68,** 939–940.
46. Roberts, J. B. G., Darlington, E. H., Edwards, R. D., and Simons, R. F. (1977). Transform coding using charge-coupled devices. *Electron. Lett.* **13,** 277–278.
47. *Proc. IEEE* **62** (5) [Special issue on SAW devices (1974)].
48. Henaff, J. (1973). Image processing using acoustic surface waves. *Electron. Lett.* **9,** 102–104.
49. Rebourg, J. C. (1978). New SAW Hadamard transformer. *IEEE Trans. Son. Ultrason.* **SU-25,** 252–256.
50. Tseng, G. C. (1972). A new transform coding and its implementation with elastic surface waves. IBM Report RC 3789, Yorktown Heights, New York.

Part Two

Applications

Chapter 5

Signal Processing

5.1 Introduction

An introduction to the use of sequency functions in signal-processing operations was given in Chapter 3, in which the special characteristics of these functions were compared with conventional Fourier methods. In this chapter attention will be directed towards a number of signal-processing applications involving these operations. As will have become apparent from earlier discussions, the major stimulus to the use of sequency functions in these applications lies in the ease with which their transformation can be implemented on the digital computer and, in some cases, the better matching of the function of the shape of the signal waveform being processed.

Signal processing in the context of this chapter is considered as those operations in which a single-dimensional stream of digital data, representing some process or time history, is manipulated to produce a simplified or improved series having a more desirable property or to estimate certain characteristic parameters of the signal. Two-dimensional processing will be considered in the following chapter. The application of sequency digital processing technique covers a very wide field and includes spectroscopy and various uses of spectral analysis; speech, seismic and other forms of low-frequency processing; radar and sonar applications; medical signal processing, particularly ECG and EEG analysis; and many others. Several of these fields have proved receptive to the application of sequency methods and will be considered here, commencing with some optical methods in spectroscopy.

5.2 Spectroscopy

One of the earliest uses of the Walsh function series was in on-line spectroscopy. Following the pioneering work of Gibbs [1] of the National Physical Laboratory, Gebbie [2] applied the Walsh transform to the decoding of the output of a two-beam interferometer. Whilst the original impetus for the work was an expectation of better matching of the functions to digital computer operations, his conclusion was that it could result in a spectroscopic system faster than that currently available and of high overall efficiency.

Some simplicity in spectroscopy equipment requirements has since been demonstrated by Despain and Vanasse [3], who used a series of Walsh-related masks to code and decode the optical signals. The use of coded masks has the effect of improving the signal-to-noise ratio of the measurement, and when the Walsh function is used, this advantage is obtained without loss of resolution.

An important development in this area was the design and marketing of an infrared spectrometer, which also uses a Walsh-coded mask, on the basis of the technique originally proposed by Decker [4] and Harwit and Sloane [5]. Two difficulties in the design of such devices were recognised: the low flux levels of infrared sources and the high attenuation at the narrow exit slit necessary to separate out the spectral elements of a dispersed light source. The better matching of the Walsh function series through the use of coded masks enabled a similar overall performance to be obtained as found in earlier scanning interferometer devices which are difficult to produce and align.

Essentially this spectrometer is a multi-slit mask device in which the collected light is passed through the orthogonally-coded mask and after detection is transformed to produce the required spectrum. A conventional dispersive element is used (fixed prism or grating) and the light distributed over the entire spectra is passed through a multi-slit mask coded in an orthogonal manner such that the slits in the mask correspond to a binary 1 and the opaque portions to a binary 0 (Fig. 5.1). The total light passing through the mask is collected on a single infrared detector. The orthogonal mask pattern is then changed and the process repeated. The process is carried out a number of times equal to the numer of discrete slots in the mask (i.e., binary 1 or 0 locations on the mask). The result is a series of output values for the detector from which the spectral values can be obtained by orthogonal transformation.

A further improvement can be made by using a double multiplexing grating. This is described by Harwit *et al.* [6]. A doubly multiplexed instrument of this type has been constructed by Phillips and Briotta [7] for use in astronomical spectral analysis.

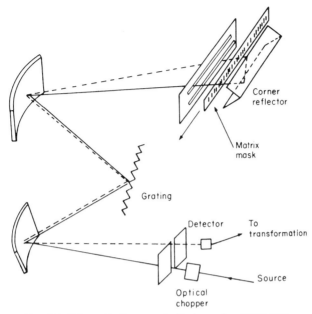
FIG. 5.1. Schematic diagram of a Hadamard spectrometer.

In quite a different area of spectral analysis the FWT has also been applied to electron-spin resonance spectroscopy to increase the speed of operations for the associated analysis [8].

5.3 Speech processing

Sequency functions are applied to the processing of speech signals in several different ways: as a method of reducing the bandwidth occupied by the transmitted speech signals, as a tool for efficient speech synthesis and as a technique for automatic speech recognition. Not all of these applications have been successful since, for the reasons stated earlier, the synthesis of sinusoidal-based signals through discontinuous waveforms is not a particularly efficient process. However, the techniques explored have been useful in the development of speech-processing methods, and the application of sequency functions for bandwidth compression has proved of lasting value in the communication field, as we shall see later.

The basic method for speech transmission is shown in Fig. 5.2. By transmitting the orthogonal transform of the speech signal rather than the signal itself, a considerable reduction in bandwidth requirements becomes possible. This is because many of the higher frequency or sequency values have

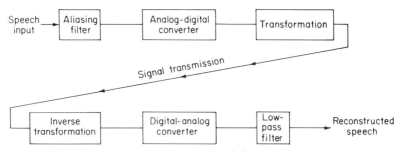

FIG. 5.2. A speech transmission system.

negligible value and can be neglected in the transmission phase without having an adverse effect on the reconstituted signal at the receiving end. By making the number of sampled values in the transmitted set assume a value of $N = 2^p$, where p is a positive integer, a fast computational algorithm such as the fast Walsh transform can be employed. Early contributions to speech bandwidth compression were made by Campanella and Robinson [9, 10] and Boesswetter [11], who showed that Walsh software coding could result in much narrower transmitted bandwidth than could conventional pulse-coded modulation (PCM) methods. Further reductions are possible if a fully hardware processing system is used, and Gethöffer [12] has described such a system in which the transmitted bit rate is half that of a PCM system.

The basic problem in speech synthesis is to maximise the efficiency of voice transmission whilst retaining acceptable distortion levels for the voice characteristics. Early work was due to Sandy [13], who first envisioned the possibilities of using a Walsh analysis for this purpose. Speech synthesis techniques have developed in terms of either a reconstruction from the power spectrum or the use of only the dominant sequency coefficients.

Gethöffer [12] has constructed a hardware system to carry out synthesis by the power spectrum method. He showed that the quantative energy distribution of German speech in the Walsh spectrum differs for voiced and unvoiced vowels. Consequently, he concludes that it should be possible to improve bandwidth compression or synthesis by adopting different coding schemes for the two voice parts. In order to adapt the time base of the coding Walsh transform, it is necessary to extract the pitch of the signal with some accuracy. Synthesis of speech and other forms of processing are then possible from the pitch-synchronous Walsh spectra produced from this adaptive system.

The use of dominant sequency coefficients in signal reconstruction is based on the assumption that the ear is insensitive to phase, so that the minimisation of phase shift obtained with the power spectrum may not be necessary. This approach has been used by Shum *et al.* [14], who have obtained fairly good reproduction of speech from a selection of only 4–8 dominant coefficients out of a total of 64 comprising the sampling set.

A number of attempts to use Walsh methods in automatic speech recognition and synthesis have been made. The problem is a difficult one, and even with a restricted vocabulary the considerable variations between different speakers lead to some complexity in the methods required for a satisfactory solution. The difficulties are due partly to the variable length for spoken identical words and, more fundamentally, to the non-stationarity of speech waveforms. This latter problem was recognised by Gethöffer [15], who devised means of continuous and parallel sequency filtering and power spectrum analysis for investigation into vowel sounds. Due to the non-stationarity of speech waveforms, it is necessary to define a fairly narrow window for the data over which stationarity can be assumed. Early work used fixed window lengths and gave poor results. Later work by Gethöffer [12] and Shum *et al.* [14] analysed the speech waveform with a window period that varies with the pitch period. An extensive discussion of these methods of synthesis and recognition is given by Flanagan [16], to which the reader is referred.

A correlation method of speech recognition is described by Clark *et al.* [17]. The method is based on the comparison between Walsh sequency components of spoken words and those taken from a stored library of words. The Walsh sequency components of successive time segments of spoken words are obtained. These components are arranged in an amplitude sequency and time matrix order and correlated sequentially against stored matrices representing the transformed values of a library of words. Recognition is obtained by extracting the word giving the highest correlation coefficient. A major difficulty lies in the variation of the duration of the spoken word mentioned earlier. This has also been noted by Edwards and Seymour [18], who compare Fourier and Walsh spectra of a restricted vocabulary of words. They conclude that, whilst the Walsh pattern contains more indications than does the Fourier pattern, this could be advantageous since small variations between different utterances of the same word have less effect on the whole pattern and should lead to more accurate identification.

5.4 Medical applications

Much of the interest in applying signal-processing technique to medical problems lies in the diagnostic importance of continuous processes within the body that can be readily monitored. Blood flow, muscular rhythms, neurological processes, respiration and heart movement are all of interest to the clinician, and electrical signals having a direct relationship to these activities are relatively easy to obtain and store. In addition to specific diagnostic measurements, there is now a requirement for continuous or on-line patent monitoring with its need for effective data compression.

Transformation techniques play a leading role in bandwidth compression and identification of significant parameters in physiological data [19, 20], and sequency methods have been applied to all of these processes. The principal applications have been in the analysis of *electroencephalographic* (EEG) and *electrocardiographic* (ECG) data, although other physiological time series have been processed through sequency methods. Walsh functions have also been applied to the modelling of biological systems. An early paper (Meltzer *et al.* [21]) described the use of Walsh functions to define a two-dimensional form, in this case morphological patterns. The study of morphogenesis is concerned with the development of an organism characterised by a succession of metabolic shapes having distinct morphological patterns. The reduction of these shapes to a series of Walsh coefficients enables the nature of the patterns and their classification to be examined with some precision.

The relationship between the Boolean function and the Walsh function has been used by Gan *et al.* [22] to determine a dynamic Boolean model of a physical system, and was developed further through the use of a related transformation in a later paper. The Walsh function has also been applied to define a system transfer function for biological systems by Seif and Gann [23] and as a filtering system in the modelling of the action of the nervous system by Boesswetter [24].

5.4.1 EEG analysis

Processing of EEG signals has been based principally on the development of time series models and spectral analysis to extract parameters of significance in the diagnosis of neural disorders and to enhance the understanding of neural processes. Perhaps the most extensive use of such methods has been in the field of sleep research [25] and in the development of automated techniques for on line monitoring [26].

Electroencephalographic signals derive from electrical activity in the cortex region of the brain and are obtained through surface-mounted disc electrodes affixed to the surface of the scalp. This is shown diagrammatically in Fig. 5.3, which indicates the pronounced difference in the activity recorded at different locations. The signals are small and associated with unwanted noise signals, and so amplification and filtering form an essential preliminary to the signal processing.

Signals obtained from brain activity are essentially sinusoidal in character and are to some extent non-stationary, so that the choice of the discontin-

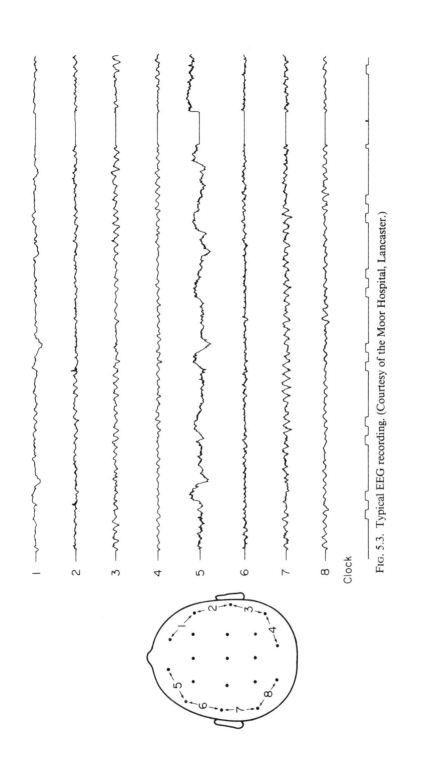

FIG. 5.3. Typical EEG recording. (Courtesy of the Moor Hospital, Lancaster.)

uous Walsh function for their spectral analysis is not an obvious one. However, there are a number of advantages in using the sequency approach which have been exploited in various ways. The most important of these is the speed advantage of applying Walsh rather than Fourier transformation, particularly in view of the large amount of data that is generally available and the need for automated techniques of analysis to reduce the magnitude of the interpretative task. When implemented on small general-purpose microcomputers which might be available for use in an EEG laboratory, the FWT is significantly faster than the FFT, and speed advantages of over 40 times have been quoted [27]. Setton and Smith [28] have also commented on the simplicity of using the Walsh approach with a small low-cost microprocessor.

There are some disadvantages, however, arising principally from the poor matching of the Walsh function waveform to the EEG signal being analysed. This shortcoming gives rise to some odd-harmonic distortion of the spectral results which does not, however, affect the central characteristic frequency (or sequency) of EEG activity (the main parameter of interest to clinicians). A discussion of the relative importance of this and other features of Walsh vis-à-vis Fourier analysis is given in [29–33].

When the periodogram definition of the power spectrum (Subsection 3.3.2) for EEG analysis is used, the effect of phase shift with the Walsh transformation is negligible [27]. This observation is confirmed by Smith [34], who also indicates that in a monitoring situation the Walsh and Fourier results provide the same information about the EEG. This arises because the normal procedure of EEG monitoring consists of consecutive time segment analyses; where the time shift from segment to segment is in the order of seconds, the segments are essentially statistically independent of each other. For some methods of analysis, such as the minimum distance clustering algorithm described by Larsen and Lai [30], the classification obtained with Walsh equals that of Fourier over the range of EEG features. The best spectral estimates were obtained by dividing a long data interval into segments, evaluating the spectra for each segment and ensemble averaging the spectra to achieve a final estimate for the entire interval. This technique can, of course, be applied with any orthogonal transformation and is a well-known procedure for non-stationary analysis [35].

As well as studies on these direct applications of sequency techniques, some work has been carried out on a *clipped* form of the EEG data. Here only the positions of the zero crossings are retained, with the positive and negative excursions of the signal replaced by values of ± 1 (Fig. 5.4). Larsen *et al.* [36] describes an application of the Walsh function to this truncated EEG data which, of course, matches the bi-level characteristic of the Walsh function and is therefore more efficient than the corresponding Fourier analysis.

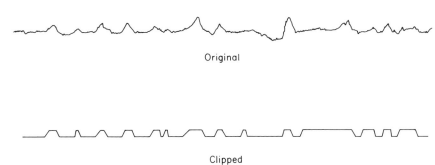

Fig. 5.4. A clipped EEG waveform.

Finally mention must be made of a hybrid technique developed by Adams [37] to overcome the problems of lack of phase invariance and poor matching of the sequency functions to the EEG waveforms. The Haar function series is modified by substituting each double-level function by single cosine and sinusoidal waveforms. The resulting Fourier–Haar functions are not quite orthogonal, and the resulting spectrum is seen to contain gaps in the continuous spectrum sequence. However, the method seems to work fairly well for the analysis of signals conforming to the *burst model* of the EEG which is accepted by many researchers. The convergence properties of these Fourier–Haar transforms are quite good and for a small number of terms are actually superior to the Fourier transform. The fast Fourier–Haar algorithm contains the simplicity of the fast Haar transform but requires additional sinusoidal weightings in some of the node calculations.

5.4.2 ECG analysis

The electrocardiogram gives a measure of the electrical activity of the heart in terms of a continuous time history. The most important characteristic of a normal cardiac cycle is the segment shown in Fig. 5.5, which repeats itself, with minor variations, once per cycle. This is known as the *QRS cycle*

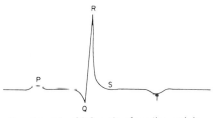

Fig. 5.5. The QRS cycle of cardiac activity.

and corresponds to the electrical activity of the ventricles during a single heartbeat. Variations in the shape of the QRS waveform are important in determining the onset of ventricular fibrillation, so that on-line analysis and classification of this segment of the cardiac system forms an important feature of automatic patient-monitoring systems. These three waveform characteristics of the cycle Q, R and S relate to particular activities of the ventricle and each has a characteristic shape which can be recognised by cardiologists. The QRS cycle is preceded and followed by smaller activity peaks P and T, which relate to atrial activity and are not as readily recognised.

For a number of reasons it is desirable to carry out this pattern recognition process automatically, and methods based on transformation instead of time domain correlation have been extensively investigated. A sequency or frequency approach to ECG analysis has a number of advantages, including the following:

1. The bandwidth compression that can be obtained allows many more patent records to be stored within a given storage media.

2. Faster identification can be obtained by using fewer significant parameters. This is important for on-line patient-monitoring systems.

3. Cases of function abnormality are more easily recognised since they are often accompanied by an enhanced higher-frequency content.

The sequency functions are very suitable for this form of analysis since the ECG function itself is discontinuous. A consistent feature of the complete waveform can easily be recognised (the QRS cycle) and used as a reference point, thus overcoming the changing variance with phase of the Walsh transformation. In addition to Walsh transformation methods, the Haar, slant and R transforms have all been applied to this recognition and classification problem.

The odd-harmonic power spectrum (Eq. (3.18)) has been used by Milne *et al.* [38] because of its powerful data reduction properties and invariance to phase shifts of the sampled ECG waveform. Since the ECG waveform is dependent upon the placement of the recording electrodes on the body, a multi-lead system is used to minimise this effect, with the electrodes placed on orthogonal axes. Signals from these two sets of electrodes are sampled simultaneously and processed as a 2×32 matrix of data points from the two acquisition channels. An analogous technique uses the Haar transform, which gives similar data reduction properties to the Walsh odd-harmonic spectrum [39]. In both cases identification of normal and abnormal cardiac conditions is made by comparison of the compressed spectral characteristics with previously obtained characteristics from measured and known cases. By using transformation values rather that a sampled time series, considerably fewer characteristics are needed for the process of comparison.

Morgan [40] has used the R transform (Section 2.6) to achieve a transform for ECG work which is insensitive to cyclic shifts of the input data. Some results of the application of simple Walsh vector filtering are also given. Here the transformation of the ECG waveform, related to the location of the QRS occurrence, is followed by selection of those components to be retained before retransformation. The resultant time domain waveform gives information related to predictive analysis of heart abnormalities such as occur prior to ventricular fibrillation.

Using the ECG signal as a basis for measurement, Thomas and Welch [41] have developed a method of heart (pulse) rate determination based directly on sequency calculation. Each time the large spike of the QRS cycle occurs, it initiates a bi-stable circuit, the state of which is sampled at a rate greater than the heart rate. The sampled values are then transformed by using a sequency-ordered Walsh transform. Under normal conditions a sampled square waveform (Rademacher function) is generated, the sequency of which gives the heart rate directly. Any variations in the triggering of the bi-stable effect that the mark: space ratio of the waveform generated will be reflected in the average zero crossing (sequency) measurement of the resulting waveform.

A statistical technique for the identification of the Q, R and S regions of the cardiac cycle is described by Poll *et al.* [42]. The ECG signal is sampled at a rate of 200 Hz, and each set of eight consecutive samples are transformed by using a fast Walsh transform algorithm. Only the first four coefficients are retained for pattern recognition purposes following normalisation to the zero coefficient. Identification is based on prior analysis of a large set of standard and normal ECG waveforms. This yields mean values of the first four Walsh transform coefficients such that a small number of simple statistical tests can be carried out on the transformed and normalised test signal. For example, in the case of the Q wave shown in Fig. 5.6, if the values of the first four coefficients W_0 to W_3 fulfil the conditions given in Table 5.1, then the Q wave is recognised.

It is necessary to move the eight-sample base progressively along the signal waveform as indicated in Fig. 5.6, with transformation for each length a, b, c, d, e to identify the best match to the Q waveform. This pattern recognition operation is extremely economical, however, requiring about 30 simple operations besides the fast transformations and is suitable for microprocessor implementation. A recognition of 97.5% for 160 random cases is claimed for the Q wave, with similar figures for R- and S-wave identification.

This technique is one of a number of statistical transformation techniques for recognition of normal and abnormal ECG waveforms which use sequency functions as described by Meffert *et al.* [43]. They also give details of signal-to-noise-ratio improvement of the signal through the use of slant spectral characteristics.

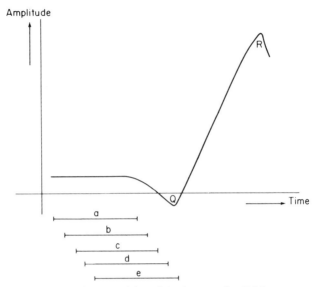

FIG. 5.6. Recognition of the Q wave of an ECG.

Excellent results in data compression for ECG data have been achieved by Ahmed et al. [44], who used the cosine and Haar transforms. Whilst the optimum transform for this purpose is shown to be the Karhunen–Loève transform, the absence of a fast transformation algorithm makes the KLT less attractive and the sub-optional transforms are preferred. The technique for compression is a simple one of discarding all transformed coefficients of higher sequency than N/M, where N is the total set of transformed values and $M:1$ the compression ratio required. A compression ratio of between $2:1$ and $4:1$ can be achieved with a MSE for the DCT quite close to that of the KLT and not significantly different for the HT. Some actual figures for $N = 128$ and various numbers of retained components N/M are shown in Table 5.2 (with acknowledgement to N. Ahmed for providing the data).

A different method of discrimination using the Haar transform is described by Hamba and Tachibara [45]. The period of the heartbeat is recognised by using differential filters and the QRS cycle recognised by using discriminant functions based on a transformation of the waveform.

TABLE 5.1. Poll et al.'s Q-wave test characteristics.

$W_0 = 1$
$W_1 > -0.075$
$W_2 > +0.0505$
$W_3 < -0.0250$

TABLE 5.2. Normalised MSEs for various transforms.

N/M	KLT	DCT	HT
8	0.2332	0.2838	0.3404
16	0.0773	0.1451	0.1763
32	0.0034	0.0293	0.0603
64	0.0000	0.0069	0.0088

The Haar function is valuable in the synthesis of pulse-like signals since only a small number of coefficients are found necessary to approximate the waveform to a small MSE (Section 1.4). This is the case with the impulsive QRS waveform, and Hamba and Tachibara show that only 60 coefficients need be retained out of a total of 2^7 coefficients to sufficiently identify the waveform. They call this process of selection a *Haar spectral filter,* and the resulting spectral values form the significant features of the waveform. The process is carried out for a number of normal and abnormal (heart-muscle-blocking) electrocardiographs to establish a method of feature identification through a linear discriminant function. Various types of heart disease can be classified into a number of separate categories. For a fixed number of terms the Haar spectral filter can specify uniquely the feature of each category, provided that at least 21 of the spectral values are considered.

5.4.3 Other medical processing

Other biological signals having quasi-periodic form have been analysed by using sequency waveforms to benefit from the simplicity or rapidity of their transformation [40, 43]. An outstanding example of this work is the application by Linkens and Cannell [46] and Temel and Linkens [47] of Walsh spectral methods to gastrointestinal signals.

These electrical signals can be recorded in many parts of the mammalian gastrointestinal tract and provide characteristic waveforms which can be related to normal or abnormal activity of the stomach, colon and duodenum and thus assist in medical diagnosis. The essential characteristics looked for are the principal rhythmic frequencies and the *spread* of the spectrum over a limited bandwidth. Although basically sinusoidal in form, the signals are obscured by signals (noise) originating from other bodily functions and are subject to distortion in their transfer to the probe electrodes. This does not affect the identification of the main frequency peaks, and the Walsh sequency-ordered spectrum proves quite adequate to detect these despite the introduction of higher-sequency subsidiary peaks since these lie outside the main spectral areas of interest.

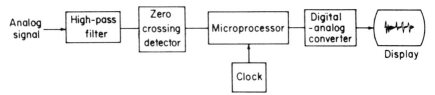

FIG. 5.7. A microprocessor system for physiological signal analysis (in Temel and Linkens [47]).

The frequencies concerned are contained within a band from 0.01 to 0.5 Hz, and on-line analysis using simple microprocessor devices is feasible.

A schematic diagram of Temel and Linkens' microprocessor system is shown in Fig. 5.7. The continuous (analog) signal is filtered to remove the very low-frequency noise accompanying the signal and sampled by a zero crossing detector. Since only the frequency of the signal is to be determined, a clipped version of the physiological waveform can be employed where a zero-crossing detector assigns a logical 1 and 0 to the positive and negative portions of the signal, respectively. The software is arranged to admit 128 values to a FWT routine carried out by the 8-bit microprocessor which, after conversion to analog form, is displayed for inspection or recording. Optional digital filtering may be carried out in the digitised data. The FWT employed is an in-place algorithm to minimise demands on memory but does require a bit-reversing routine carried out by software after transformation. Consecutive sets of 128 values taken from the sampled analog signal can be processed in this way or a refreshing software routine used to maintain the spectrum of the initial set of data on the display device.

The system has been used effectively as part of clinical diagnosis to differentiate visually the spectrum of normal and abnormal colon operation and to indicate the effects of drug absorption.

5.5 Seismology

Many of the advances in understanding the mechanisms of seismic wave transmission and determining the origins of seismic disturbances have been based on the use of Fourier transformation and spectral analysis records, or *seismograms*. A number of attempts have been made to use other transformations, particularly the Walsh transform, and some useful results have been obtained. The motivation has been the saving in computational time or memory requirements and also in recent years the simplicity and speed of transformation using the microprocessor.

In an early paper by Båth and Burman [48] Walsh spectral analysis is applied to the detection of the vertical components of Rayleigh waves initiated by underground nuclear explosions and the results compared with similar analysis using the Fourier transform. Despite the essentially sinusoidal nature of the seismic signals, Båth and Burman conclude that, since the Walsh spectrum essentially characterises the signal, it is of value as a simply performed operation to get a general indication of the spectral nature of the signal.

An interesting feature of the Walsh spectrum for seismic events is that the spectral energy in the sequency domain appears to fall into two groups, shown as M and N in Fig. 5.8(a). The primary group M contains most of the energy and corresponds to the Fourier spectrum shown in 5.8(b). The secondary group N contains the higher sequency components not especially present in the Fourier representation. This effect has also been noted by Kennett [49], who suggests that this secondary group can be useful as an aid to earthquake pattern recognition.

The presence of these secondary signals can be understood if we consider a derivation of the Walsh power spectrum from the dyadic convolution of a time series analogous to the Wiener–Khinchine derivation with Fourier analysis. We see from Eq. (3.5) that dyadic convolution requires modulo-2 addition of the delayed signal. For most series such a correlation will be a mixture of samples having both long and short delays (Fig. 3.1), and the resulting spectrum will require higher-order Walsh components to describe them.

A further difference noted between the use of Fourier and Walsh transforms lies in the derivation of true ground motion from the derived spectrum by correcting for the response characteristic of the seismic transducer. The ground motion can be obtained in the Fourier case from the relationship between the Fourier transform of the convolution of two signals as being equal to the product of the Fourier transforms of the two signals. As we saw earlier, such a simple relationship between convolution and the product of two signals does not exist with the Walsh function; this situation makes it difficult to remove the effect of the transducer characteristics from the derived signal.

A comparison between the use of Fourier and Walsh spectral analysis of marine seismic data has been given by Chen [50] and Chen and Boucher [51]. Here in addition to the computational saving obtained by the use of the FWT over the FFT, a non-recursive filtering is carried out in the sequency domain by using linear and non-linear Wiener filters. A considerable improvement in signal-to-noise ratio is obtained in the Walsh case for the examples given by Chen, who describes improvements in seismic feature extraction over the use of purely frequency methods.

Seismic signals contain non-stationary characteristics, and within an individual seismogram there is considerable variation in the distribution fre-

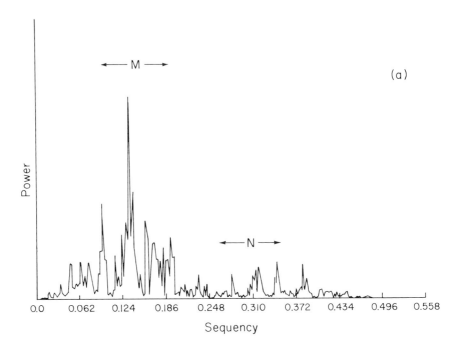

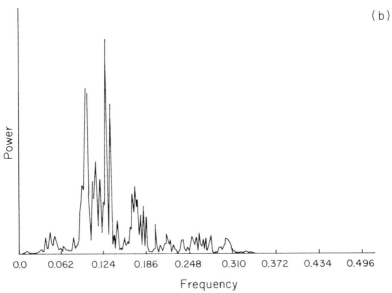

FIG. 5.8. Seismic even-power spectrum (a) using the Walsh transform and (b) using the Fourier transform.

quency energy found in different parts of the record. Hence there is some interest in spectral display methods which plot spectral energy against time. Kennett [49] has carried this out by using Walsh transformation and has compared the results for different events and locations with similar results from short-term Fourier spectra. An extra complexity in the sequency – time plots is produced for the Walsh spectra. This takes the form of higher-sequency *side-bands* which can indicate interesting similarities between apparently dissimilar records. Examples are given by Kennett to illustrate a convenient diagnostic characteristic for the seismogram which enables event classification to be made.

A current trend in seismological measurement is the use of distributed arrays of seismometers often extending in two or more directions over an area of several square kilometers. The signals derived from individual seismometers in the array are combined in various ways either to improve the detection capability in terms of improved signal-to-noise ratio or to permit determination of the direction of arrival for the seismic disturbance. Early attempts to improve plane wave detection with a mixed Fourier – Walsh spatial transform were not very successful, although the use of the Walsh power spectrum reduced the amount of computation required [52]. This finding was due to the presence of interfering side-bands generated by the Walsh analysis. Whilst this prevents reasonable approximation to the actual received frequency spectrum, the sensitivity obtained with array detection is enhanced by the use of fast Walsh transforms. One consequence is that attention has now shifted from using Walsh methods in the reconstruction of the seismic plane wave, which can be carried out effectively by Fourier means, to the solution of the signal detection problem in which the speed of the FWT is paramount.

Significant advances have been made recently in *array detection capability* since the publication by Goforth and Herrin [53] of an automatic signal detection algorithm using the Walsh transform. Since digital seismic data are obtained as a time series of amplitudes, many detection systems use the amplitude of the signal as the basis for a decision. However, the waveforms from seismic events differ from background noise not only in amplitude but also in frequency content. Hence the interest in using a detector based on a transform of the data where changes in amplitude and frequency can be sensed and used for determining the occurrence of seismic events. Whilst the FFT can be used in a detection algorithm, its requirement for many floating-point multiplications makes it too slow for real-time applications when a small microprocessor is being used, and all the applications of the Goforth and Herrin algorithm have used the FWT as a basis for the calculations.

The detection algorithm is shown in Fig. 5.9 (with acknowledgements to Drs. Goforth and Herrin). The seismic signal is digitised in real time (i.e.,

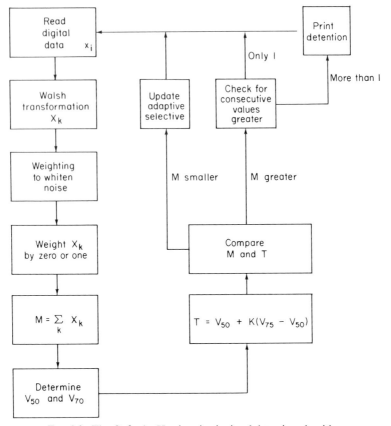

FIG. 5.9. The Goforth–Herrin seismic signal detection algorithm.

continuously) with processing being carried out on a *window* of 64 samples (3.2 s) of data which are updated by 32 samples as each new set of samples is processed so that an overlap of detection and processing can occur. This procedure implies, of course, that the processing for event detection must be capable of operating faster than real time. The current set of 64 samples is Walsh transformed and the transformed coefficients X_k weighted in such a way as to *whiten* the Walsh spectrum of the noise. The weights are factors such as $\frac{1}{8}, \frac{1}{4}, \frac{1}{2}, \frac{3}{4}$ and 1 which can be applied by shifts and fixed-point addition. In some versions of the algorithm the coefficients are further weighted by 0 or 1 to isolate the expected sequency band of the signal (see Chapter 3). Event detection is obtained by comparing the median value of the samples which are summed to give an M value of $M = \Sigma_k X_k$ with an adapted threshold value.

The absolute values of the coefficients, rather than their squares, are

5.5 Seismology

summed since this appears to give a more stable parameter in this application and is much faster to compute. The detection threshold, which is computed from the distribution of the previous 512 sums of absolute values, is defined by

$$T = V_{50} + K(V_{75} - V_{50}) \qquad (5.1)$$

where T is the threshold, V_{50} is the median of the distribution of the previous 512 values of M, V_{75} equals 75% of the distribution of the previous set of 512 values and K is an arbitrary constant. If the current value of the sum of the absolute values of the Walsh coefficients M exceeds the threshold T, a signal is called. If it does not exceed the threshold, the sum of the absolute values is ranked among the previous 512 values, the oldest value being discarded. In this way an adaptive detection threshold is maintained with an adaptation window (in this case) of approximately 14 min. If a signal is called, the threshold is not adapted.

In practical terms an adaptive threshold means that the threshold follows a running average of the signal with a time lag. If the signal is increasing, then an increase in threshold level occurs and an event is indicated if the signal input continues to rise. If the signal decreases, the average value follows it down, so that small significant events can still be detected.

The algorithm has been implemented at several sites in the United States, including the New England MIT seismic network [54], and in Norway on the NORSAR seismic array [55]. Results have been uniformly good at all sites and equivalent or better than those obtained from experienced geoscientists working on the visual record. In a typical real-time on-line test extending over several months at the NORSAR array, the algorithm was able to detect 1347 events with a false alarm rate of 0.21 per hour. This compares with 1041 events and a false alarm rate of 0.70/hr when the amplitude detection NORSAR event detection system was used over the same period.

In most of these systems the algorithm is implemented in machine language by using an 8-bit microprocessor of the Z80A type. A block diagram of Goforth and Herrin's [53] system is shown in Fig. 5.9. Three channels of short-period (20 samples per second) data are analysed in real time, with a logical flow diagram essentially as shown in Fig. 5.10. The system has available 48K bytes of RAM storage and with this capability is able to process up to 10 streams of data in real time.

The flow diagram for the MIT system is similar but with the squares of the Walsh sequency CAL and SAL coefficients added to yield a power sequency spectrum which is unaffected by phase shifts [54]. A faster 16-bit processor is used which permits this and other enhancements to be implemented without affecting real-time operation.

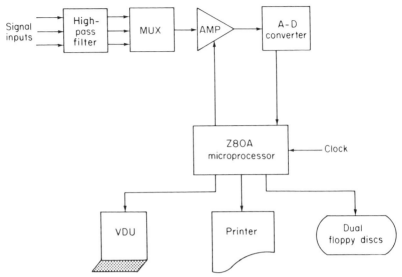

FIG. 5.10. A microprocessor implementation.

5.6 Non-linear applications

An interesting area of application which makes use of the unique features of the Walsh series is that of non-linear stochastic problems. In particular, some success has been obtained in improving the efficiency of signal detection for those transducers which are essentially non-linear in operation.

The special feature of Walsh series found useful in this connection is illustrated in Fig. 5.11. If we assume that a given waveform can be represented by a group of Walsh functions of a given order, the result will be a stair-step approximation to the waveform. If this is then operated upon by a single-valued non-linear transformation, then another stair-step function is obtained which has the same number of steps but with changed step heights. Any further non-linear operations will have a similar result such that the amplitudes of the individual steps will change but not their total number. This has the effect of limiting the number of new sequency terms generated. From Eq. (2.18), we see that the only intermodulation products produced by the multiplication of two Walsh functions will be the modulo-2 addition of each of the sequency terms. Therefore, if the input signal can be represented by a finite number of Walsh functions N, then only

$$(N-1)(N-2)(N-3) \cdots 1$$

intermodulation products will be generated. The situation is different with the Fourier representation of the input signal. Here a set of new harmonics would be generated which will have combination frequencies equal to the

5.6 Non-Linear Applications

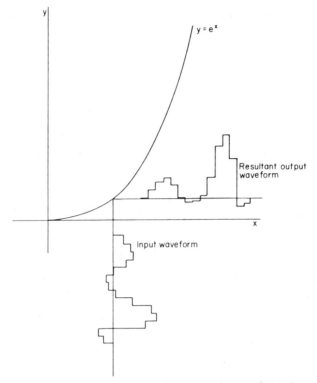

FIG. 5.11. Non-linear operation on a stair-stepped signal.

sum differences of all the possible harmonic values of the signal waveform and the non-linear function.

An example will illustrate the difference between these two types of representation, both operated on by the same type of non-linearity. If we define a signal to consist of two Walsh functions, viz.,

$$x(t) = \text{WAL}(r, t) + \text{WAL}(s, t) \tag{5.2}$$

which are passed through a power-law device having an output of the form

$$y(t) = a_1 x(t) + a_2 x^2(t) + a_3 x^3(t) \tag{5.3}$$

then $y(t)$ is given as

$$\begin{aligned} y(t) = {} & a_1[\text{WAL}(r, t) + \text{WAL}(s, t)] + a_2[\text{WAL}^2(r, t) + \text{WAL}^2(s, t) \\ & + 2\,\text{WAL}(r, t)\,\text{WAL}(s, t)] + a_3[\text{WAL}^3(r, t) + \text{WAL}^2(s, t)\,\text{WAL}(r, t) \\ & + 2\,\text{WAL}^2(r, t)\,\text{WAL}(s, t) + \text{WAL}(s, t)\,\text{WAL}^2(r, t) \\ & + \text{WAL}^3(s, t) + 2\,\text{WAL}(r, t)\,\text{WAL}^2(s, t)] \end{aligned} \tag{5.4}$$

Using the addition theorem for the Walsh transform (Eq. (2.17)),

$$\begin{aligned}y(t) &= a_1[\text{WAL}(r, t) + \text{WAL}(s, t)] \\ &\quad + a_2[2\,\text{WAL}(r \oplus s, t) + 2\,\text{WAL}(0, t)] \\ &\quad + a_3[\text{WAL}(r, t) + \text{WAL}(r, t) + 2\,\text{WAL}(s, t) \\ &\quad\quad + \text{WAL}(s, t) + \text{WAL}(s, t) + 2\,\text{WAL}(r, t)] \\ &= 2a_2\,\text{WAL}(0, t) + (a_1 + 4a_3)\,\text{WAL}(r, t) \\ &\quad + (a_1 + 4a_3)\,\text{WAL}(s, t) + 2a_2\,\text{WAL}(r \oplus s, t) \end{aligned} \quad (5.5)$$

Thus, the output consists of the desired terms having changed amplitude coefficients plus a dc term and an intermodulation product.

If we now define the input signal as consisting of two sinusoidal functions, viz.,

$$x(t) = \sin A + \sin B \quad (5.6)$$

where $A = f(\omega r t)$ and $B = f(\omega s t)$, then, if $x(t)$ is passed through a device having the relationship given by Eq. (5.3), the output signal would be

$$\begin{aligned}y(t) &= a_1[\sin A + \sin B] + a_2[\sin^2 A + \sin^2 B + 2 \sin A \sin B] \\ &\quad + a_3[\sin^3 A + \sin^3 B + \sin A \sin^2 B + \sin B \sin^2 A \\ &\quad\quad + 2 \sin^2 A \sin B + 2 \sin^2 B \sin A] \\ &= a_1[\sin A + \sin B] + a_2[\tfrac{1}{2}(1 - \cos 2A) + \tfrac{1}{2}(1 - \cos 2B) \\ &\quad + \cos(A - B) - \cos(A + B)] \\ &\quad + a_3[\tfrac{1}{4}(2 \sin A - \sin 3A - \sin(-A)) + \tfrac{1}{4}(2 \sin B - \sin 3B - \sin(-B)) \\ &\quad + \tfrac{3}{2}(2 \sin A - \sin(A + 2B) - \sin(A - 2B)) \\ &\quad + \tfrac{3}{2}(2 \sin B - \sin(B - 2A) - \sin(B - 2A))] \end{aligned} \quad (5.7)$$

Here the output consists of the original signal plus its second and third harmonics, a dc term and six intermodulation products. The position is much more complex and careful bandpass filtering is necessary to extract the required signal from the complex modulated output. In the general case for Walsh functions only high- or low-pass sequency filtering is needed.

Several workers have considered the effects of non-linearities, as described above, on the Walsh modulation of a sinusoidal carrier signal. With any multi-channel carrier system of communication of this type, non-linearities within the system will give rise to cross-modulation products (crosstalk), which can be serious. These products can be discriminated more easily by using Walsh modulation compared with sinusoidal modulation. However, in the case of Walsh functions, there is a danger that the modulo-2 addition of the addition theorem will result in Walsh functions being generated which will coincide with one or more of the desired signals. To avoid this, the use of a Rademacher subset of the Walsh functions has been suggested by Frank and Harmuth [56]. Rademacher functions form an incomplete set and have the property that their products yield a Walsh function that cannot be a Rademacher function. This system has the disadvantage in a practical case that wider transmission bandwidth would be required. Harmuth [57] suggests other alternative methods of selection for the modulating signals to minimise the cross-talk without incurring this penalty.

Corrington [58] gives several examples related to the non-linear modulation operations involved in phase-shift keying. He shows that the derivation of the frequency spectrum, through the use of Walsh functions, considerably eases the analytical problems involved. As a further example, we can consider the application of Moss [59], who has investigated the use of a Walsh series to modify a pseudo-random binary sequence used in the estimation of the impulse response of a non-linear system (gas chromatography).

The input transducer for such a system consists of an on–off sampling valve connected to a chromatographic column. The gas detector connected to this column is frequently highly non-linear, so that the estimation of linear impulse response would normally have a large variance. Since the input system can switch between only two levels (the input valve on–off positions), the linearisation techniques proposed prior to Moss's work have been only partially successful in removing the effects of the square and cubic terms. The Walsh method employs only two levels of input and is effective in such a situation.

The modified pseudo-random binary sequence is applied as an input to the system, i.e., operating the sampling valve. Operation of the non-linear detector on the admitted gas samples results in the generation of an output sequence modified by the transfer function of the gas-chromatographic column. In measuring the cross correlation between the output due to the modified pseudo-random binary sequence and the sequence itself, a function results which contains contributions from the even-power terms of the non-linearity only. Hence, as long as the only even power present is in the second (a function of this particular system), the method is applicable with equal success in situations in which the output non-linearity contains an arbitrary number of higher-order odd-power terms.

References

1. Gibbs, J. E. (1970). Walsh spectroscopy: A form of spectral analysis well-suited to binary digital representation. National Physical Laboratory report, Teddington, England.

2. Gebbie, H. A. (1970). Walsh functions and the experimental spectroscopist. *Proc. Symp. Applic. Walsh Functions, Washington, D.C.*, AD707431, pp. 99–100.
3. Despain, A. M., and Vanasse, G. A. (1972). Walsh functions in grille spectroscopy. *Proc. Symp. Applic. Walsh Functions, Washington, D.C.*, AD744650, pp. 30–35.
4. Decker, J. A. (1972). Hadamard transform spectrometry: A new analytical technique. *Anal. Chem.* **44**, 127–134.
5. Harwit, M., and Sloane, J. A. (1979). "Hadamard Transform Optics." Academic Press, New York.
6. Harwit, M., Phillips, P. G., Fino, T., and Sloane, J. A. (1970). Double multiplexed dispersive spectrometers. *Appl. Opt.* **9**, 149–154.
7. Phillips, P. G., and Briotta, D. A. (1974). Hadamard-transform spectrometry of the atmospheres of Earth and Jupiter. *Appl. Opt.* **10**, 2233–2235.
8. Evans, J.C., and Morgan, P. H. (1981). Automatic analysis of mixed spectra. *Analyt. Chim. Acta* **133**, 329–338.
9. Campanella, S. J., and Robinson, G. S. (1970). Digital sequency decomposition of voice signals. *Proc. Symp. Applic. Walsh Functions, Washington, D.C.*, AD707431, pp. 230–237.
10. Campanella, S. J., and Robinson, G. S. (1971). A comparison of Walsh and Fourier transformations for application to speech. *Proc. Symp. Applic. Walsh Functions, Washington, D.C.*, AD727000, pp. 199–205.
11. Boesswetter, C. (1970). Analog sequency analysis and synthesis of voice signals. *Proc. Symp. Applic. Walsh Functions, Washington, D.C.*, AD707431, pp. 220–229.
12. Gethöffer, H. (1972). Speech processing with Walsh functions. *Proc. Symp. Applic. Walsh Functions, Washington, D.C.*, AD744650, pp. 163–168.
13. Sandy, G. F. (1969). Speculations on possible applications of Walsh functions. *Proc. Symp. Applic. Walsh Functions, Washington, D.C.*, pp. 387–415.
14. Shum, Y. Y., Elliott, A. R., and Brown, W. O. (1973). Speech processing with Walsh–Hadamard transforms. *IEEE Trans. Audio Electroacoust.* **AU21**, 174–179.
15. Gethöffer, H. (1971). Sequency analysis using correlation and convolution. *Proc. Symp. Applic. Walsh Functions, Washington, D.C.*, AD727000, pp. 119–123.
16. Flanagan, J. L. (1972). "Speech Analysis, Synthesis and Perception." Springer-Verlag, Berlin.
17. Clark, M. T., Sanson, J. E., and Sanders, J. A. (1972). Word recognition by means of Walsh transforms. *Proc. Symp. Applic. Walsh Functions, Washington, D.C.*, AD744650, pp. 169–172.
18. Edwards, I. M., and Seymour, J. (1973). Discrete Walsh functions and speech recognition. *Proc. Theory Applic. Walsh Functions, Hatfield Polytechnic, England.*
19. Bonner, R. E. (1970). Electrocardiogram monitoring by computer, *in* "Clinical Electrocardiography and Computer" (C. A. Caceres and L. S. Dreifus, eds.). Academic Press, New York.
20. Start, L., Okjima, M., and Whipple, C. H. (1962). Computer pattern recognition techniques. *Commun. Assoc. Comput. Mach.* **5**, 10.
21. Meltzer, B., Searle, N. H., and Brown, R. (1967). Numerical specification of biological form. *Nature* **216**, 32–36.
22. Gann, D. S., Seif, F. J., and Schoeffler, J. D. (1972). A quantized variable approach to description of biological and medical systems. *Proc. Symp. Applic. Walsh Functions, Washington, D.C.*, AD744650, pp. 134–141.
23. Seif, F. J., and Gann, D. S. (1972). An orthogonal transform approach to the description of biological and medical systems. *Proc. Symp. Applic. Walsh Functions, Washington, D.C.*, AD744650, pp. 128–133.

24. Boesswetter, C. (1972). Modelling the compound action potential of the nerve. *Proc. Symp. Applic. Walsh Functions, Washington, D.C.,* AD744650, pp. 142–149.
25. Chase, M. H. (1972). "The Sleeping Brain: Perspectives in the Brain Sciences," Vol. 1. Brain Res. Inst., Los Angeles.
26. Gervius, A. S., Yeager, C. L., Diamond, S. L., Spine, J. P., Zeithin, G. M., and Gevins, A. H. (1975). Automated analysis of the electrical activity of the human brain: A progress report. *Proc. IEEE* **63**, 1382.
27. Weide, B., Andrews, L. T., and Iaunone, A. M. (1978). Real time analysis of EEG using Walsh transforms. *Comput. Biol. Med.* **8**, 255–263.
28. Setton, J. J., and Smith, W. D. (1979). An EEG monitor using the Walsh transform on a standard microprocessor. *IEEE Trans. Biomed. Eng.* **BME-26**, 525.
29. Yeo, W. C., and Smith, J. R. (1972). Walsh power spectra of human electroencephalograms. *Proc. Symp. Applic. Walsh Functions, Washington, D.C.,* AD744650, pp. 159–162.
30. Larsen, H., and Lai, D. C. (1980). Walsh spectral estimates with applications to the classification of E.E.G. signals. *IEEE Trans. Biomed. Eng.* **BME-27**, 485–492.
31. Jansen, B. H. (1981). Comments on Walsh spectral estimates with applications to the classification of E.E.G. signal. *IEEE Trans. Biomed. Eng.* **BME-28**, 667–668.
32. Larsen, H., and Loi, D. C. (1981). Authors reply [to [31]]. *IEEE Trans. Biomed. Eng.* **BME-28**, 668.
33. Jansen, B. H., Bourne, J. R., and Ward, J. W. (1981). Spectral decomposition of E.E.G. intervals using Walsh and Fourier transforms. *IEEE Trans. Biomed. Eng.* **BME-28**, 836–838.
34. Smith, W. D. (1981). Walsh versus Fourier estimators of the E.E.G. power spectrum. *IEEE Trans. Biomed. Eng.* **BME-28**, 790–793.
35. Beauchamp, K. G. (1973). "Signal Processing Using Analog and Digital Techniques." Geo. Allen & Unwin, London.
36. Larsen, R. D., Crawford, E. F., and Howard, G. K. (1976). Walsh analysis with E.E.G. signals. *Math Biosci.* **31**, 237–253.
37. Adams, E. R. (1977). Non-stationary time series: The Fourier–Haar transform. *IEEE Conf. Random Signal Anal. Pub.* **159**, 36–52.
38. Milne, P. J., Ahmed, N., Gallagher, R. R., and Harris, S. G. (1972). An application of Walsh functions to the monitoring of electrocardiograph signals. *Proc. Symp. Applic. Walsh Functions, Washington, D.C.,* AD744650, pp. 149–153.
39. Ahmed, N., and Rao, K. R. (1974). Data compression using orthogonal transforms. *Proc. Symp. Applic. Walsh Functions, Washington, D.C.,* pp. 236–245.
40. Morgan, D. G. (1971). The use of Walsh functions in the analysis of physiological signals. *Proc. Theory Applic. Walsh Functions, Hatfield Polytechnic, England.*
41. Thomas, C. W., and Welch, A. J. (1972). Heart rate representation using Walsh functions. *Proc. Symp. Applic. Walsh Functions, Washington, D.C.,* AD744650, pp. 154–158.
42. Poll, R., Stoschek, E., Henssge, R., and Kaiser, K. (1979). Anwendung der Walsh-transformation zur Bestimmung von Einzelmustern am Elektrokardiogram. *Z. Elektr. Inform u Energietechnik, Leipzig* **9**, 305–316.
43. Meffert, B., Schubert, D., Lazarus, T., Poll, R., and Hensage, R. (1980). New and known methods of the application of transforms to quasiperiodic biomedical signals. *IEEE Symp. Electromag. Compat. Baltimore,* pp. 336–341.
44. Ahmed, N., Milne, P. J., and Harris, S. G. (1975). Electrocardiograph data compression via orthogonal transforms. *IEEE Trans. Biomed. Eng.* **BME-22**, 484–487.
45. Hamba, S., and Tachibara, Y. (1976). Representation and recognition of electrocardiograms using finite Haar transform. *Elect. Eng. Jpn.* **96**, 111–117.

46. Linkens, D. A., and Cannell, A. E. (1974). Interactive graphic analysis of gastro-intestinal electrical signals. *IEEE Trans.* **BME-2,** 335–339.
47. Temel, Z. B., and Linkens, D. A. (1978). Medical data analysis using microprocessor-based Walsh transformations. *IEEE Trans. Biomed. Eng.* **BME-16,** 188–194.
48. Båth, M., and Burman, S. (1972). Walsh spectroscopy of Rayleigh waves caused by underground explosions. *Proc. Symp. Applic. Walsh Functions, Washington, D.C.,* AD744650, pp. 48–63.
49. Kennett, B. L. N. (1974). Shot-term spectral analysis and sequency filtering of seismic data, *in* "Exploitation of Seismograph Networks" (K. G. Beauchamp, ed.), pp. 283–296. Noordhoff, Leiden, Netherlands.
50. Chen, C. H. (1972). Walsh domain processing of marine seismic data. *Proc. Symp. Applic. Walsh Functions, Washington, D.C.,* AD744650, pp. 64–67.
51. Chen, C. H., and Boucher, R. E. (1973). Further results on Walsh domain processing of marine seismic data. *Proc. Symp. Applic. Walsh Functions, Washington, D. C.,* AD763000, pp. 253–256.
52. Lintz, P. R. (1973). Walsh function detection and estimation of plane waves at an array of seismometers. *Proc. Symp. Applic. Walsh Functions, Washington, D.C.,* AD763000, pp. 248–252.
53. Goforth, T., and Herrin, E. (1981). An automatic seismic signal detection algorithm based on the Walsh transform. *Bull. Seism. Soc. Am.* **71,** 1351–1360.
54. Michael, J., Gildea, S. P., and Pulli, J. J. (1982). A real-time digital seismic event detection and recording system for network application. *Bull. Seism. Soc. Am.* **72,** 2339–2348.
55. Veith, K. F. (1981). Seismic signal detection algorithms. Teledyne Geotech, Garland, NTIS Report AD-A110 186/4, Washington, D.C.
56. Frank, T. H., and Harmuth, H. F. (1971). Multiplexing of digital signals for time-division channels by means of Walsh functions. *Proc. Theory Applic. Walsh Functions, Hatfield Polytechnic, England.*
57. Harmuth, H. F. (1969). Applications of Walsh functions in communications. *IEEE Spectrum* **6** (11), 82–91.
58. Corrington, M. A. (1962). Advanced analytical and signal processing techniques. Rome Air Development Center, Griffiths Air Force Base, New York. NTIS Report AD277942, Washington, D.C.
59. Moss, G. C. (1971). The use of Walsh functions in identification of systems with output nonlinearities. *Proc. Theory Applic. Walsh Functions, Hatfield Polytechnic, England.*

Chapter 6

Image Processing

6.1 Introduction

The processing of fixed or changing visual images by using digital techniques requires the manipulation of multi-dimensional signals involving operations on large numbers of data values. Since this process generally involves high computer costs and substantial processing time, strenuous attempts have been made to find efficient solutions to the problems involved. The sequency techniques, with their emphasis on rapid computational development, have been found to play a significant role in these developments, and their use will be explored in this chapter.

Examples of multi-dimensional signals may be found in facsimile transmission, television images, medical x-ray photographs, radar and sonar maps, satellite-transmitted pictures, electron-microscope records, seismic data displays and many other applications. For most purposes these signals consist of black-and-white pictures which are subsequently sampled and quantised to give a series of digitised *picture elements* (pels) which are actually digital numbers expressing a brightness value over the picture area. This digital form of a picture will be termed an image or image matrix.

Processing these data can take many forms. It is convenient here to consider three particular areas which among them cover nearly all the expected processing requirements:

(a) *Image compression* a reduction in the transmission channel or digital storage requirements for an image by means of efficient coding schemes.

(b) *Image enhancement* the improvement of picture quality in some way or the reduction of some of the degradations incurred in the acquisition or transmission of the image.

(c) *Pattern recognition* the detection and extraction of particular patterns or features from an image for the purpose of classification or for facilitating easier recognition of some identifiable feature.

In this chapter emphasis will be placed on discrete sequency transformation, often using the natural-ordered Walsh or Hadamard transform, as an essential means of data manipulation. Other techniques based on optical methods, statistical operations or various forms of pulse encoding fall outside the scope of this book and are fully discussed elsewhere [1–3].

An introduction to the two-dimensional transform was given in Section 2.8, where it was shown that the computational process comprises two operations: a transformation carried out on the rows of the image followed by a second transformation of the columns of the result to achieve a two-dimensional transformation with a single-dimensional transform. It is instructional to consider this transformation in matrix terms since such can enable the appreciation of a physical meaning for this transformation.

A two-dimensional Walsh function expansion is shown in Fig. 6.1. It consists of the first 8×8 functions of the Walsh series arranged in sequency order in which each rectangle represents a different combination of line and column image definition. We can consider a two-dimensional image matrix x_{ij} to be transformed into its Walsh domain $X_{m,n}$ through a process of matrix multiplication and summation by each of the rectangular functions shown in Fig. 6.1. Thus,

$$X_{m,n} = x_{1 \cdot 1} W_{1 \cdot 1} + \cdots + x_{ij} W_{ij} + \cdots + x_{8 \cdot 8} W_{8 \cdot 8} \quad (6.1)$$

viz.,

$$X_{m,n} = \sum_{i=1}^{N} \sum_{j=1}^{N} x_{ij} \cdot W_{ij} \quad (6.2)$$

(neglecting normalisation by $1/N^2$).

From these equations we can see that for a given image representation of a picture the values of the two-dimensional coefficients obtained will represent a correlation, or 'match', of the entire picture area with each of these Walsh representations. Thus a transformation of the chequerboard pattern shown in Fig. 6.2 will result in a large coefficient for the $W_{5 \cdot 4}$ term and zero value for the other coefficients. This matching property forms the basis of several feature extraction algorithms and is discussed further in Section 6.4.

The number of pels defined in the picture-sampling process limits the definition of the reconstructed picture, and attempts to process a picture

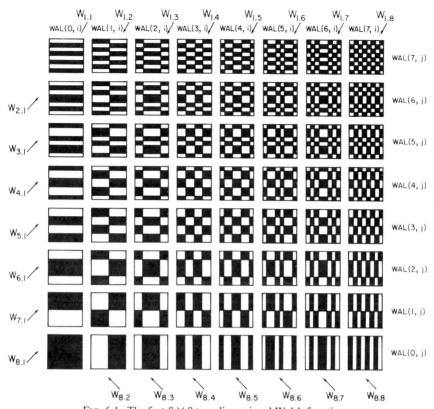

FIG. 6.1. The first 8 × 8 two-dimensional Walsh functions.

having inherently greater definition will effectively insert a sequency-filtering process ahead of the sampling and quantisation. In general quantisation levels are selected to minimise subjective errors in the reconstructed picture, and although a linear set of levels are usually chosen, there are advantages in Gaussian or logarithmic form [1].

Another image-processing characteristic of which we should be aware is the method of processing through sub-picture elements. The picture is divided into rectangular sub-pictures, and these are sampled and quantised to form rectangular groups of pels for subsequent processing or transmission. This process introduces the possibility of parallel or pipeline processing of

FIG. 6.2. A chequerboard pattern.

these sub-matrix images and can improve coding efficiency [4]. The choice of sub-matrix size is dependent on the type of transformation selected, and for n^2 pels will be found to lie between $n = 4$ and $n = 16$. Whilst the MSE should improve with increasing n as the number of correlations taken into account also increases, it has been found that most pictures contain significant correlation between adjacent pels, and no substantial improvement in performance is found for $n > 8$ [4, 5].

6.2 Image compression

The efficient coding for data transmission or storage of an image to reduce channel or storage capacity is termed *image compression*. What is sought here is a reduction in the number of digital bits describing the image such that fewer bits need to be stored or transmitted over the digital communication system. Examples of areas in which this technique proves useful include television or videophone transmission, nationwide computer and time-sharing networks, satellite image transmission and the efficient storage of x-ray records for hospitals.

Two general approaches to the problem have been investigated. One uses spatial domain techniques, such as intraframe coding techniques utilising the considerable redundancy in adjacent picture frames [6], and the other uses transformation techniques. This second approach will be considered in some detail, commencing with the problem of choice of transformation method.

6.2.1 Choice of transform

A number of orthogonal transformations have been employed for image compression, including the Fourier, sine, cosine, Hadamard, slant, Haar and Karhunen–Loève. As noted previously the Karhunen–Loève is the most efficient in terms of mean-square error [4] but requires N^3 arithmetic operations for an $N \times N$ image and no fast algorithmic procedure seems possible. It is, however, taken as a reference for compressional performance, although other transforms are preferred in practical applications. The sequency transforms all offer fast transformations requiring $2N^2 \log_2 N$ additions/subtractions or fewer without the complex value arithmetic and storage penalty of the Fourier transform. The Hadamard [7, 8], slant [9, 10] and Haar [11, 12] transforms have all been used for this purpose.

The Haar transform is the fastest of the three, requiring $2N(N - 1)$ simple arithmetic operations, but it has a poor decorrelation performance compared with the other two. The slant transform has been applied specifically to television transmission because of its good match with the characteristics of the scanned image (see Section 2.5). It provides a considerable improvement in image compression over the spatial domain methods using pulse-coded modulation and is considerably faster than Fourier transformation methods. Until the recent advent of fast cosine transformation, the Haar transform's MSE performance proved superior to those of other fast transformations and its computational efficiency only a little poorer than that of the Hadamard transform. The superior energy compaction property of the cosine transform may be seen from a plot (Fig. 6.3) of comparative MSE performances of the various transformations as a function of sub-matrix size [9]. Its comparison with the ideal KLT can also be seen in the diagram.

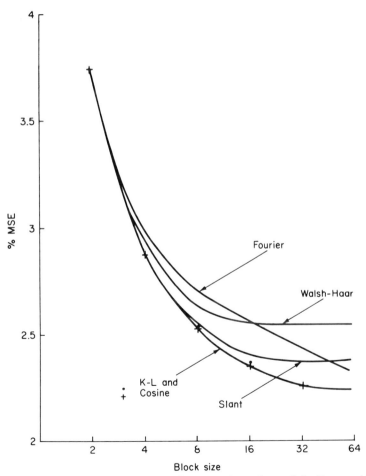

FIG. 6.3. Mean-square error performance of orthogonal transforms. (After Pratt *et al.* [9].)

The two factors looked for in image coding are compressional efficiency and ease of computation. At present the cosine transform provides the best approach to these two factors. Its MSE is nearly equivalent to the KLT and a number of fast implementations are available.

The fast transformations were initially implemented by using the FFT as a basic algorithm and complex arithmetic [13–15]. This approach involves computation of a double-size FFT of $2N$ coefficients employing complex arithmetic. An alternative algorithm, known as a fast discrete cosine transform (FCT), requiring only real operations is also available and is approximately six times as fast as the FFT implementation [16]. This can be used with fast sequency transforms, and the WHT algorithm has been applied in several recent implementations of the fast cosine transform.

Two improved methods using the Hadamard matrix are described by Hein and Ahmed [17] and Ghanbari and Pearson [18]. These methods are realised through a conversion matrix having a block diagonal structure and follow the general transform conversion method described by Jones et al. [19] and discussed in Section 2.7.

Somewhat simpler hardware construction is required for a particular approximation to the cosine transform, known as the C matrix transform and described by Srinivassan and Rao [20], which has the advantage that the conversion matrix has only integers as its elements.

A sine transform having similar capabilities for image processing has also been defined, but it has less efficient compressional characteristics [21]. It can yield a faster computational algorithm than the DCT. A useful review of its performance and implementation is given by Yip and Rao [22]. A method of deriving this transform from the cosine transform is described by Wang [23].

Several of these transforms for image processing will be considered next, commencing with the discrete cosine transform.

6.2.2 The discrete cosine transform

The discrete cosine transform pair may be defined as

$$X_{ct}(n) = \frac{1}{N} \sum_{i=0}^{N-1} x_i \cos \frac{n\pi(2i+1)}{2N} \qquad (6.3)$$

and

$$x_i = \sum_{n=1}^{N-1} X_{ct}(n) \cos \frac{n\pi(2i+1)}{2N} + X_{ct}(0) \qquad (6.4)$$

where $X_{ct}(0)$ is the average value $(1/N)\sum_{i=0}^{N-1} x_i$ and $i = 0, 1, \ldots, N-1$, $m = 0, 1, \ldots, N-1$. Equation (6.3) may be recognised as having a close relationship with the continuous cospec function [24] and can be derived directly from the complex Fourier transform (Equation (2.9)) as

$$X_{ct}(n) = \text{Re}\left\{\exp\left(-\frac{jn\pi}{2N}\right) \sum_{i=0}^{2N-1} x_i \exp\left(-\frac{j2\pi ni}{N}\right)\right\} \quad (6.5)$$

(neglecting scaling), where $j = \sqrt{-1}$ and $\text{Re}\{\cdot\}$ denotes the real part of the term enclosed. From this relationship it is easy to see that the FFT forms a conventional route to the DCT. This was originally described by Ahmed *et al.* [5] in terms of a FFT whose data length was doubled by adding zeros. An alternative realisation by Schaming [14] obtained a similar result by using the FFT output plus its mirror image. In either case, a double-length structure was found necessary to permit simple extraction of the real values of the complex result shown in Eq. (6.5). A slightly faster derivation requiring less data storage is described by Haralick [15] and applies the FFT to two single-length data vectors of the original data.

The value of the cosine transformation for image compression lies in its good variation distribution and low rate of distortion function. This results in efficient energy compaction where the transform coefficients containing the largest variances are found to be contained in approximately a quarter of the transformed matrix. This energy compaction is virtually the same energy compaction found with the KLT, known to be optimal in the mean-square sense. Its variance characteristics also lend themselves to efficient adaptive coding compression techniques which are based on the statistics of the cosine transformed image. In the scheme described by Chen and Smith [25], transformed sub-pictures are sorted into classes by the level of image activity present. Within each activity coding, bits are allocated to individual transform elements according to the variance matrix of the transformed data. Bits are then distributed between the high-activity (busy) and low-activity (quiet) areas, with few bits allocated to the quiet areas. A bit rate of 1 bit/pel for a monochrome image is obtained and double this for a colour image.

6.2.3 The fast cosine transform

A flow diagram for the FCT for $N = 8$ is given in Fig. 6.4. The diagram consists of alternating cosine–sine butterflies with binary matrices to reorder the matrix elements to a form which preserves a bit-reversed pattern at every other node. It is thus an in-place algorithm and apart from normalisation can be reversible and used as an inverse transform. Extension of the flow diagram to the next power of two involves adding a set of ±1 butterflies and a series of alternating cosine–sine butterflies to yield a new set of odd

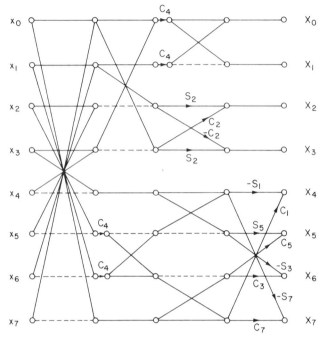

FIG. 6.4. A flow diagram for the FCT. [$C_i = \cos(i\pi/16)$, $S_i = \sin(i\pi/16)$.]

transform coefficients. In matrix terms the transform matrix for the FCT can be written as

$$\mathbf{CT}_N = \mathbf{P}_N \left[\begin{array}{c|c} \mathbf{CT}_{N/2} & \mathbf{O} \\ \hline \mathbf{O} & \mathbf{R}_{N/2} \end{array} \right] \mathbf{I}_N \qquad (6.6)$$

where $\mathbf{P}_N$ is an $N \times N$ permutation matrix translating the transformed vector from bit-reversed to natural order and $\mathbf{I}_N$ is an identity matrix. The sparse matrix containing the previous order $\mathbf{CT}_{N/2}$ and a complex matrix $\mathbf{R}_{N/2}$ permits extension into a higher order $\mathbf{CT}_N$ with suitable algorithmic decomposition of $\mathbf{R}_{N/2}$. Chen and Smith [25] show that $\mathbf{R}_{N/2}$ consists of four distinct types of sine–cosine matrix, each of which can be decomposed to result in an algorithmic operation to produce $\mathbf{CT}_N$. The result is about six times faster than the double FFT requiring some $3(N/2)(\log_2 N - 1) + 2$ real additions and $N \log_2 N - 3(N/2) + 4$ real multiplications [16]. The implementation of the direct algorithm is still a complex procedure and one that is difficult to carry out effectively in hardware terms for real-time image compression unless a highly modular parallel processing approach is taken, with consequent increase in control complexity.

An alternative is to obtain the FCT via the Walsh transform, which reduces the number of non-integer multiplications required.

6.2 Image Compression

If we consider the DCT matrix for $N = 8$ shown below, we see that a recognisable symmetry exists in the signs of the cosine matrix terms, viz.,

$$\mathbf{CT}_8 = \begin{bmatrix} 0.354 & 0.354 & 0.354 & 0.354 & 0.354 & 0.354 & 0.354 & 0.354 \\ 0.490 & 0.416 & 0.278 & 0.098 & -0.098 & -0.278 & -0.416 & -0.490 \\ 0.462 & 0.191 & -0.191 & -0.462 & -0.462 & -0.191 & 0.191 & 0.462 \\ 0.416 & 0.098 & -0.490 & -0.278 & 0.278 & 0.490 & -0.098 & -0.416 \\ 0.354 & -0.354 & -0.354 & 0.354 & 0.354 & -0.354 & -0.354 & 0.354 \\ 0.278 & -0.490 & -0.098 & 0.416 & -0.416 & 0.098 & 0.490 & -0.278 \\ 0.191 & -0.462 & 0.462 & -0.191 & -0.191 & 0.462 & -0.462 & 0.191 \\ 0.098 & -0.278 & 0.416 & -0.490 & 0.490 & -0.416 & 0.278 & -0.098 \end{bmatrix}$$

(6.7)

There is, in fact, a one-to-one correspondence between the signs of these terms and those of the Walsh matrix $\mathbf{W}_8$ shown in Eq. (1.48). The correspondence indicates that the basic vectors of the DCT are essentially *amplitude-modulated* versions of the basic vectors of the FWT, and this provides a rationale for the development of a conversion algorithm between the two transformations.

It has been shown elsewhere [17] that the link between $\mathbf{W}_N$ and $\mathbf{CT}_N$ takes the form of a transformation matrix having a diagonal block structure containing many zero value terms. This is shown below for $N = 8$.

$$\mathbf{D} = \begin{bmatrix} 1.0 & & & & & & & \\ & 1.0 & & & & & & \\ & & 0.923 & 0.383 & & 0 & & \\ & & -0.383 & 0.923 & & & & \\ & & & & 0.907 & -0.075 & 0.375 & 0.180 \\ & 0 & & & 0.214 & 0.768 & -0.513 & 0.318 \\ & & & & -0.318 & 0.513 & 0.768 & 0.214 \\ & & & & -0.180 & -0.375 & -0.075 & 0.907 \end{bmatrix}$$

(6.8)

Fewer multiplications are required to implement this matrix compared with those required for Eq. (6.7), and, despite the need first to compute the FWT, the number of additions–subtractions is also reduced.

A similar method is described by Ghanbari and Pearson [18] and applied to the implementation of a television image compression system. In order to reduce the number of arithmetic elements needed in this implementation, the Hadamard ordering (e.g., $\mathbf{H}_8$ of Eq. (1.47)) is taken as the sign-comparative Walsh transform and the columns and rows of the DCT matrix rearranged to conform with this vector basis. The advantage of this is that with a suitable decomposition algorithm the adder and subtractor and associated delay circuits can be shared between all the samples in the 8 × 8 pel and a single shift in place of the input data register produces all four pairs of data for subsequent arithmetic operation. This type of hardware-sharing operation was discussed earlier in Chapter 4. The matrix form of the resulting FCT for $N = 8$ is given as

$$CT_8 = \{D_1 + D_2\}H_2 - H_3 \cdot x_i \qquad (6.9)$$

where H_2 and H_3 are Hadamard matrices and D_1 and D_2 represent sparse block-structured matrices similar to Eq. (6.8) but containing far fewer non-zero terms. The order of the input and output terms, however, is not sequential and the input terms x_i will need to be arranged in dyadic order and the output coefficients in bit-reversed order. A complete hardware image compression system using this technique is described by Ghanbari and Pearson [18] and Ghanbari [26]. It is designed to operate a 625-line monochrome television system at an 11 MHz sampling rate and satisfactory performance obtained at a reduction of the original density of 8 bits/pel down to a transmission rate of 2.4 bits/pel.

6.2.4 Image transmission

The transmission of still and moving images is a major application for image compression. Until fairly recently the transmission of still images has meant facsimile transmission in which image reconstruction can take several minutes and no advantage is realised in using compressional techniques. However, with the growing use of videotext the transmission of still images is now required in seconds. Because of the slow transmission rates presently in use, it is essential that some form of image compression be used.

The transmission of moving images (e.g., television) is much faster, and, although a wider bandwidth is allowed, there are still considerable advantages in using data compression. Actually, in terms of the product of bandwidth and reconstruction time the two cases are very similar. Presently, videotext uses a transmission rate of 4800 bits/s or less with about 6 s for reconstruction giving a product of approximately 30,000 bits/s s. Television transmission of 625-line standard requires a frame reconstruction time of 1/50 s and, assuming 8-bit coding to provide full luminance and colour gradation, the product is approximately 60,000 bits/s s. Hence we find that similar image compression methods are used in both of these applications.

Transform coding for videotext (picture Prestel) transmission using Hadamard coding has been described by Nicol *et al.* [27]. Here only one-ninth of the screen area is used for picture information, the remainder allocated to alphanumeric information, which is much less demanding in data transmission rate and storage [28]. The system described is arranged to encode and transmit single colour television frames as an insert picture. The colour signal is carried by separate luminance and colour difference signals with 4 bits allocated for each luminance and chrominance pel. In the experimental British Telecom system a Hadamard transformation is carried out on the image prior to coding for transmission. The picture is divided into sub-pictures of 8×8 pels which are transformed separately. Since the distribution of energy over the television spectrum is non-uniform, with most of the energy in the lower part of the video spectrum, it is possible to truncate the actual coefficients transmitted and so to achieve data compression. By transmitting these significant coefficients in an hierarchical rather than a block-by-block basis, a gradual increase in picture resolution can be achieved. In addition, since the coefficients for each block arrive one at a time, decoding involves a simple arithmetic operation, whereas the conventional approach is to wait for all the coefficients to arrive for each block before inverse transforming. Thus a fast transformation algorithm is not needed. The effect of transmitting the lower coefficients from each block first is to produce a complete, but low-definition, picture insert. Subsequent sets of coefficients add to the stored picture in the receiver as the higher-sequency coefficients are transmitted. Figure 6.5 illustrates the effect of progressive picture generation using this technique with four out of the possible eight images shown.

At the low rates of transmission presently used for this system, namely, 1200 bits/s, it is possible to carry out the Hadamard inverse transformation directly by using an 8-bit microprocessor in the receiver (Intel 8080). At 4800 bits/s a faster 16-bit machine of the Intel 8086 type would be necessary.

As noted earlier, data compression techniques for moving television images are successful because of the high level of correlation that occurs between the gray levels of spatially adjacent picture elements. This provides the redundancy that can be exploited in intra-frame coding, of which transform coding forms one important technique [29]. The reduction in transmitted coefficients is obtained by threshold elimination in which all those coefficients which fall below a specified value are not retained and only those higher than this level transmitted. Other methods of selection are used and described in the literature [25, 29]. Another source of redundancy that can be removed relates to the high correlation that exists between subsequent frames of the transmitted television picture. This is known as inter-frame coding. Sequency techniques have been applied to both intra-frame and inter-frame coding.

A data compression technique for intra-frame colour television transmission has been described by Jalali and Rao [30]. It also sub-divides the picture into a number of sub-images of 4, 8 or 16 pels for transformation and transmission of the significant coefficients. Here the DCT is used, and, by

FIG. 6.5. A gradual image build-up using an 8 × 1 Hadamard transform. (a) First image, (b) second image, (c) fourth image, (d) eighth image. (From Nicol *et al.* [27].)

adopting a highly modular structure in a pipeline configuration, it has been possible to maintain a throughput rate of the NTSC carrier at three times the sub-carrier frequency of 10.7 MHz. A similar flow diagram for the FCT (Fig. 6.4) is used in this implementation but with the arithmetic operations carried out in a pipeline structure. As the input data block is collected at the input terminals, the preceding data block is being processed by the first stage of butterflies, which delivers its output to be stored at the input of the next set whilst the data previously stored there are being processed simultaneously. The process is similar to that described in Chapter 4, with similar advantages and limitations. To simplify hardware operation three different butterfly logic configurations are used: a basic butterfly involving addition and subtraction only, a similar butterfly involving sine and cosine multiplication at the nodes and a composite butterfly involving three basic butterfly operations. These are shown as (a), (b) and (c) in Fig. 6.6 and may be identified in the flow diagram of Fig. 6.4. The construction of the complete transform and its control are given by Jalali and Rao [30].

Most inter-frame compression systems use two-dimensional transform

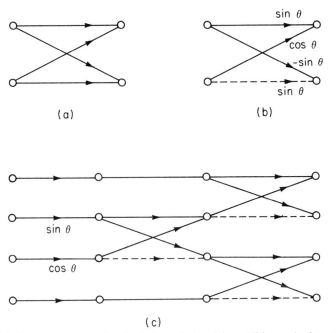

FIG. 6.6. Butterfly modules: (a) a basic butterfly involving addition and subtraction; (b) a butterfly involving sine and cosine multiplication; (c) a composite butterfly.

coding, often utilising a cosine transform in conjunction with differential pulse-code modulation (DPCM) between successive frames. This technique encodes the difference between a given transform coefficient and the corresponding encoded coefficient from the preceding scan line for the frames of an interlaced scanning system. There is usually built in to the DPCM feedback loop an attenuation factor that diminishes the value accumulated from previous entries prior to its combination with a new value. This practice provides some immunity against transmission error by including a self-correcting feature for synchronisation faults. A survey of such systems is given by Kamangar and Rao [31] in which Hadamard, cosine and Fourier transformations are compared.

An early paper in which the use of a Hadamard transform is used is Heller's [32], and complete systems have been described by Jones [33, 34]. In these papers an operational black-and-white television encoder is described which has been used in a satellite communication linkage between Stanford and Carleton Universities in the United States. Each sampled picture element is digitised to 8 bits, giving a data rate of 64M bits/s. This is reduced to 2 bits/pel by taking a 4 × 4 sub-group of picture elements and performing a fast Hadamard transformation and quantising the coefficients in a non-linear manner [35]. This rate is further reduced to 1 bit/pel by adding memory to the encoder and decoder process and transmitting only the differences of the Hadamard coefficients between successive frames. This procedure attains a final transmission rate of 8M bits/s, thus achieving a compression ratio of 8 : 1. A later development uses an adaptive coding to adjust the MSE of the inter-frame coding, dependent on the correlation measured between adjacent frames which considerably improves the performance for moving pictures [34].

A similar method is employed in the DPCM techniques described by Kamangar and Rao [31] for colour television transmissions. The technique uses a hybrid coding scheme in which transform coding is used for intra-frame coding, followed by a predictive quantisation of the inter-frame data [36].

6.3 Image enhancement and restoration

It is an unfortunate fact that image data is rarely representative only of the image of interest but is generally accompanied with other undesirable information. We may wish to abstract the image from its background, remove random noise, often inherent in the means of image acquisition, and correct various optical or other aberrations such as motion blur or defocussing. The techniques used for these purposes are called image *enhancement* and *restoration.*

Enhancement is designed to improve image quality for human viewing and brings a certain amount of subjective evaluation into the process. Restoration is the reconstruction of the image by inversion of some degradation phenomena, and for this to be successful some form of knowledge concerning the degradation characteristics will be required. This may take the form of analytic, statistical models or other a priori information based on the known information about the image structure. We will find also that optical signal images have the special property of always being positive quantities that therefore do not obey completely normal statistical operations [37]. Image restoration may, in fact, be considered as an *ill-conditioned process* and as such lacks a unique solution. For this reason we find a number of empirical methods in use, although some broad trends in image treatment are discernible. The restoration process often is not a linear one, and it is common to find that non-linear methods produce superior results than do linear ones.

6.3.1 Edge detection

Because subject outlines or edges represent a powerful description of images, detecting edges is an important preliminary process in many image-analysis situations. By *detection* is usually meant the enhancement of contrast ratio such that only the outline of the subject remains. Detection is often followed by a shape recognition process, which is considered later in this chapter.

Several methods have been studied for automatic digital edge extraction. Three of the most common are

(a) two-dimensional frequency or sequency filtering,
(b) local operations through gradient detection and
(c) contrast enhancement by transformation.

An edge of an image corresponds to high-frequency or high-sequency components in the X and Y directions. A high-pass or bandpass digital filter may be applied to the image and a contrast expansion obtained. This technique does not perform well on noisy images, however, since the noise itself consists mainly of higher-frequency components. The use of a Walsh function series in the sequency filtering has the advantage of a function matching more closely the characteristics of the edge discontinuity, and good results have been obtained by Wang [38] through the use of a circular sequency filter [39].

Local operators detect edges and contours by extracting gradients through a test on a given picture element and its neighbouring pels. This is usually associated with a threshold value, so that if the gradient difference exceeds the threshold it is assumed that at that point that there is an edge whose direction is orthogonal to the gradient direction. Another method of gradient detection is to employ a set of *templates,* or masks, of different orientation to search sequentially at each point for the best match between the selected sub-image and the mask [40]. This method is applied by O'Gorman [41] to the identification of edge direction through the use of two-dimensional Walsh functions. Here the matching is simplified by approximating both the template and picture sub-image with truncated orthogonal expansions. By arranging that the sub-image vector is chosen to be a power of two, the boundaries of these sub-images will coincide with the rectangular grid form of a two-dimensional Walsh representation (see Fig. 6.1). In this way, one form of mismatch due to digitisation of the image can be avoided. The actual comparison of the sub-images with an ideal edge template needs to take into account the position and orientation of the edge present in the sub-image. This is carried out by selecting a limited number of two-dimensional Walsh functions to represent the sub-images and applying these coefficients in an analytical calculation involving a set of template coefficients. These latter coefficients relate to edge orientation, and an MSE test reveals which sub-image coefficients are to be retained in the image reconstruction.

With transform methods it is easy to see that sequency functions can be significant owing to the *match* of these functions to the abrupt discontinuity looked for. This is particularly the case with the Haar function and its double-step characteristic. All currently used transforms are global in concept since they involve processing all the pels in a sub-image. There is no inherent selection of the high spatial sequency components which contain edge representation. The Haar transform, on the other hand, has the majority of its basis functions representing well-defined edges in the spatial domain as edges in the transform domain; i.e., it is a local rather than a global transform. Figure 6.7 shows the localised nature of the two-dimensional Haar function series (compare this with the equivalent Walsh series shown in Fig. 6.1).

Because of this local processing characteristic, the Haar transform is ideal to use in conjunction with thresholding or selection of the n largest coefficients. This process has the effect of an adaptive compression of the coefficients in which the bit compression is relatively low in areas of high image activity (at the picture contours) and high in other areas. Examples of the improvement in image compression for the Haar transform in this application are given by Lynch and Reis [42], who show that for edge representation in a 4×4 image array the Haar transform requires at most 10 coefficients for negligible error, whereas the global transforms generally require all 16 coefficients.

A considerable improvement in edge contrast for the Haar transform has also been noted by Sivak [43, 44], who applies threshold selection to infrared images. In this application the four highest-spatial-sequency coefficients out of a 4×4 matrix are retained and others reduced to 25% of their original value. First the horizontal lines of the digital image are transformed through a single-dimensional FHT and subject to a threshold operation to enhance

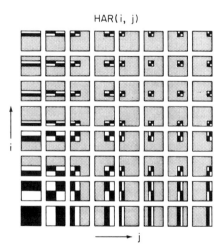

FIG. 6.7. Localised nature of the two-dimensional Haar function series.

6.3 Image Enhancement and Restoration

vertical edges. Followed by similar operations on the vertical lines to enhance horizontal edges, this process also improves the signal-to-noise ratio for the image.

A Haar-like series of two-dimensional functions, proposed by Shore [45], comprises three types of orthonormal two-dimensional functions derived from products of a Haar series. The first of these has the normal double-step of the Haar function S_n^{ij}, the second has a horizontal shape H_n^{ij} and the third a vertical shape V_n^{ij}. These are illustrated in Fig. 6.8 for the initial functions of the series. A partial sum $P_N(x, y)$, which defines the transformation, is then given by

$$P_N(x, y) = Co + \sum_{n=1}^{N} \sum_{i=1}^{2^{n-1}} \sum_{j=1}^{2^{n-1}} \left[a_n^{ij} S_n^{ij}(x, y) + b_n^{ij} H_n^{ij}(x, y) + c_n^{ij} V_n^{ij}(x, y) \right]$$
(6.10)

where Co is a constant term. The coefficients a_n^{ij}, b_n^{ij} and c_n^{ij} constitute the two-dimensional transformation. It is suggested that this type of transformation is appropriate for image transmission where large areas of the scene are constant or slowly changing with time. It is also useful for edge detection since the coefficients b_n^{ij} are sensitive to horizontal edges whilst c_n^{ij} are sensitive to vertical edges of the object outline [46].

6.3.2 Image enhancement

A major objective in image enhancement is the improvement in signal-to-noise ratio. In order to determine some quantitative values for this improvement, Kennedy [47] has carried out Walsh filtering of a given quan-

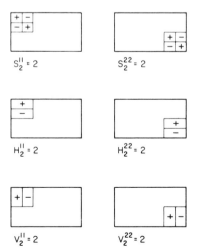

FIG. 6.8. Haar-like two-dimensional functions. (After Shore [45].)

tised and sampled image subjected to various amounts of additive noise. Under certain conditions, he has found that the addition of up to 50% of additive noise to a picture transmitted in sequency form has little discernible effect on the reconstructed picture.

The design of optimum filters for image enhancement has proceeded by the development of various forms of matched filtering. Since it is very inconvenient to have to know the position of the desired feature contained within the data, alternative sub-optimum solutions have been proposed. One of these, due to Treitel and Robinson [48], is to derive the filter transform coefficients not directly from the required signal but from another signal that has the same power spectrum as the required signal. The only requirement is that for real data the filter output must also be real; i.e., its transfer function must be symmetric. With the use of a Fourier transformation the filter constants are obtained by taking the modulus of the complex Fourier transform of a matching image having the desired size, shape and orientation. It is not quite so easy to do this with a Walsh transform since the result is not positionally invariant. Instead, the Walsh power spectrum is taken which combines the squares of CAL and SAL functions to give approximate independence from the effects of phase shift, which would be very apparent if the transform were used directly. A similar spectrum is obtained from the sum of the squares of the Haar transform, but in this case it is necessary to average the absolute value of all the coefficients derived from functions of the same degree.

Gubbins *et al.* [49] have applied this form of matched filtering to data simulating magnetic measurements made on buried archaeological sites. The structures encountered are usually of geometric shape and therefore readily identifiable, but due to irregularities in the upper soil layers, the strength of the magnetic anomalies is low and subject to poor signal-to-noise ratios. The Walsh filtering used gives good results with much less computational time than equivalent Fourier filtering. Haar filtering is also described as giving results which are less acceptable although obtained in a shorter time.

A second fairly common enhancement technique is the method of grey-level transformation which is much used for x-ray-image contrast improvement. The method is a straightforward application of Wiener filtering (Fig. 3.10). Two-dimensional transformation of the image is taken first along the rows of the digitised image and then along the columns of the resulting data as described earlier. Modification of the transformed coefficients is then carried out. This can be simply the preferential weighting of the higher-sequency terms to enhance the outlines of the image or more complex methods involving non-linear or inverse Gaussian filtering. Finally, inverse transformation of the weighted coefficients yields the enhanced digital image. Examples are shown by Andrews [50] and Pratt [51] and by Hall and Kahveci [52], who applied this method to x-ray image enhancement.

6.3.3 Image restoration

Image degradation caused by the data collection environment (e.g., mist, cloud or rain) or by the characteristics of the optical system used (blurring, defocussing, etc.) may be corrected through the use of an *inverse filtering technique* [53, 54].

If the total system has a frequency transfer function $\tau(\omega)$, then an inverse filter may be defined as a filter having a response of

$$F(\omega) = 1/\tau(\omega) = \tau^{-1}(\omega) \qquad (6.11)$$

The situation is illustrated in Fig. 6.9, from which it is clear that the output $o(x, y)$ of the inverse filter will be theoretically equivalent to $i(x, y)$—the true input image. In a practical case noise will be introduced at N and the reconstructed image in the frequency domain will be

$$o(\omega) = (i(\omega)/\tau(\omega)) - (N(\omega)/\tau(\omega)) \quad \text{for} \quad \tau(\omega) \neq 0 \qquad (6.12)$$

If the noise is large, separate filtering is required to reduce the effect of $N(\omega)$ and to simplify the process so that as far as possible the output image before retransformation approximates to

$$o(\omega) \simeq i(\omega)/\tau(\omega) \qquad (6.13)$$

Where the signal-to-noise ratio is high over the image bandwidth, inverse filtering can yield quite good results [55].

The problem with this method is to obtain a value for the system transfer function, referred to as the point-spread function (PSF), and this may be difficult if a priori information on the system characteristics is less than complete. Wiener filter restoration requires extensive a priori information on the point-spread function so that its transform may be computed for the filter [51]. A technique for generating adaptive two-dimensional low-pass filters which uses the Walsh transform and which does not depend on a priori knowledge is described by Smith [56], but it is only applicable to reducing the

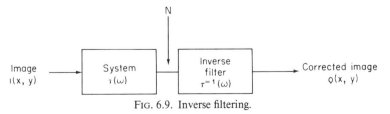

FIG. 6.9. Inverse filtering.

image noise. Here image filtering is implemented by using a procedure designed to eliminate statistically insignificant Walsh power spectra before inverse transforming of the data to restore the modified image.

Other methods not dependent on a priori information include a minimum MSE method, entropy and homomorphic techniques, but a discussion of these and the ways of deriving an acceptable PSF for a given image fall outside the scope of this book. (See Cappellini *et al.* [39] and Hunt [57].)

6.4 Pattern recognition

A problem in pattern recognition is to generate descriptions of images such that they can be related to models classifying a given set of images or patterns. A broader view of the identification problem exists in which sets of general processes are matched against the data, but in this book the narrower view will be taken and only image pattern recognition will be discussed. Many different mathematical techniques are used for pattern recognition. Two general approaches are the *decision theoretic* or statistical approach in which we look for characteristic measurements, called features, in the image and attempt to classify from these measurements; and the *syntactic* or linguistic approach in which a hierarchical information structure is produced [58, 59]. In the latter case, we may classify the pattern in terms of simpler sub-patterns and each simpler sub-pattern can again be described in terms of even simpler sub-patterns, and so on.

The two techniques are illustrated in Fig. 6.10. In the decision theoretic case the problem splits into extraction of the image features followed by pattern classification on the basis of these features. Matching of the pattern with a set of feature values stored in a memory, referred to as *template matching,* is one important method of pattern classification.

6.4.1 Decision theoretic approach

Most of the applications for sequency functions fall into this group. The general approach is to subject the image data to a two-dimensional transformation and attempt classification based on the image transform coefficients. In many cases each class of pattern will have a few dominant transform coefficients, and this should allow high probability of correct classification. An approach towards ensuring few dominant characteristics is to choose a transform with characteristics matching the pattern class of interest. Clearly this approach will be ineffective if the basic features of each class differ widely. However, if the classes have certain features in common, this will be a valid technique. The two-dimensional Walsh functions, shown in terms of

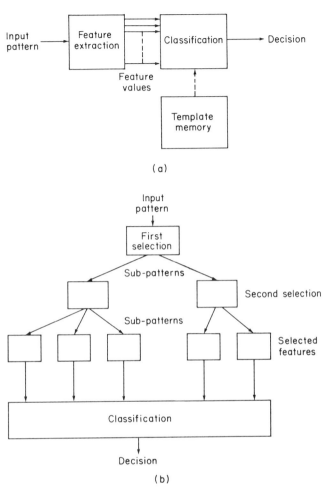

FIG. 6.10. (a) Pattern recognition: the decision theoretic approach; (b) pattern recognition: the syntactic approach.

Cartesian co-ordinates in Fig. 6.1, should therefore be applicable in detecting straight lines, rectangles and squares contained within the image. This technique has proven to be the case, and good results are obtained by using Walsh functions to match the patterns observed from man-made structures, e.g., buildings, dams, roads and urban construction. It may also be noted that a two-dimensional form of the Walsh function exists, based on polar co-ordinates which should be most adept in detecting circles, discs or rings [60].

Many investigators have applied sequency methods to the identification of alphanumeric characters either handwritten or machine-printed materials [61-64]. In most of these a sequency feature set is obtained through domain transformation and applied to some form of *minimum-distance classifier* or compared with an empirical set of spectra for template matching.

A minimum distance classifier uses the distances between the input pattern and a set of reference values in the same domain. Let m reference values $R_1, R_2, \ldots, R_n$ be associated with a given class C_i. One minimum distance classification scheme with respect to a given R_i is to classify the input as class C_i when

$$X = C_i \quad \text{if} \quad |X - R_i| \text{ is a minimum} \quad (6.14)$$

where $|X - R_i|$ is the distance defined between X and R_i. We can, of course, define $X - R_i$ in the sequency domain, and this is often done because of the smaller number of features found necessary to describe a given class C_i. It can be shown that a minimum-distance classifier is also a linear classifier [59].

One method of template matching is to determine the correlation between the image or its transform and the image of a set of arbitrary pattern descriptors [61, 65] and to accept that the data matches the reference template when a given normalised correlation value is exceeded.

A more complex example of written character recognition lies in the recognition of signatures and the detection of forgeries. Nemcek and Liu [66] describe several Hadamard transformation and configurations used to achieve these results. The signatures are presented to the computer by using a television camera input system; presentation is followed by a digitised and size-normalisation procedure. The transformation of the data is followed by a clustering transform which clusters the members of a given class in a second transform domain by minimising the mean-squared distance between samples [67]. A verification stage also computes a minimum distance factor from the mean value of the class to which the signature is assumed to belong and which has been previously stored in the computer memory.

It is generally accepted that the use of sequency functions in feature extraction for numerical data is most effective in terms of efficiency and speed due to the functions' matching capability and the use of fast transform algorithms. However, there is a general problem in dealing with the positional variance of these functions [68]. In most cases the Walsh power spectrum is used instead of its transform (see Section 2.6) to obtain a reasonably non-invariant characteristic. The Haar power spectrum and the non-invariant R transform have also been used [63, 65, 68]. Sequency methods have been utilised for other forms of shape identification as well; examples are found in ECG characteristic identification [69], in mass-spectroscopic analysis [70] and more recently in positional control for robotics [71].

The most successful applications have been in the identification and classification of rectangular structures, and an interesting area is that of topographical identification through the use of the Walsh transform, which is described by Chen and Seemuller [72] and others. A summary of this work follows as an example of the application of sequency methods applied to the decision theoretic approach.

The techniques discussed are applied to the recognition in aerial photographs of straightline roads, road intersections and rectangular buildings, but they could be used in similar feature extraction situations. The transformed data (or rather a subset of them) are subject to an algorithm carrying out a selection and classification procedure to recognise a set of selected cartographic features from this digitised and transformed set of image coefficients.

The gray-shade distribution of selected topographic features is first translated into an analog array signal and processed in a threshold operation to obtain a two-dimensional binary representation of the image. This threshold filtering is an effective first step to remove background scene information and noise. A two-dimensional Walsh transformation of this binary image provides a matrix spectral signature of the topographic features for subsequent computer analysis. A block schematic diagram of the method is shown in Fig. 6.11 [73]. A photographic transparency of the region of interest is held in a viewing stage illuminated with a white light source and projected on to a 32 × 32-element solid-state area sensor. The resulting analog array signal is digitised and after threshold filtering is stored as a two-dimensional binary array of 32 × 32 pels representing the spatial signature of the selected topographic feature. A two-dimensional Walsh sequency transformation of the stored array is carried out through repeated row and column single-dimensional fast transforms (Section 2.8). A much faster device for direct optical

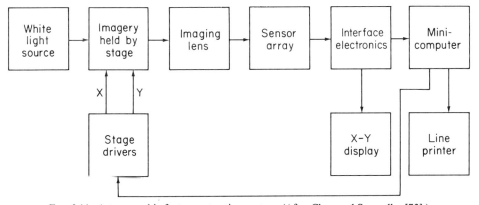

FIG. 6.11. A topographic feature extraction system. (After Chen and Seemuller [73].)

Walsh transformation that uses a rectangular plasma discharge mechanism has also been used for this application and is described by Chen *et al.* [74].

The transformed data are then subject to a series of classification tests to determine the characteristic topographic feature contained in the spatial region examined. The detection and classification scheme is based on the uniqueness of the Walsh transform of each feature under examination. A reference signal is established for each cartographic feature of the entire selected set, and these signals are compared sequentially in the Walsh domain with the transformed test image to enable classification to be carried out. This process is essentially a correlation between the characteristic *shape* of the Walsh functions and the rectilinear patterns sought in the spatial image.

An elementary example would be the identification of a horizontal straightline road which yields large coefficients in the first transformed matrix column (Fig. 6.12a). Similarly, a vertical straightline road gives significant coefficients in the first transformed matrix row (Fig. 6.12b).

In order to classify accurately topographic features it is necessary for these to be recognisable from any angle. Further the features may appear in a variety of locations within the viewing window (the active surface of the image sensor). Because the Walsh transformation is neither translationally nor rotationally invariant, two or more classification tests are required for each class of cartographic features to avoid misclassification. The tests themselves, however, are considerably simplified due to the spectral compression that occurs with most images, limiting consideration to only a small number of low-order Walsh coefficients containing the significant values. (We meet this useful characteristic again in Chapter 8 for a quite different form of spectral evaluation.)

The reference signal signatures consist of the magnitudes of a single Walsh coefficient or the sum of a row or column of transform coefficients and in the ratio of each significant coefficient to certain key low-order coefficients dependent on the feature to be determined. For example, to differentiate between straightline roads and intersections (two major groups of selected features), the first 16 rows and 16 columns of the transform matrix are summarised as

$$C_j = \sum_{i=1}^{16} |X(i,j)|, \qquad R_i = \sum_{j=1}^{16} |X(i,j)| \qquad (6.15)$$

where $X(i,j)$ is the ith row and jth column element of the Walsh coefficient matrix. These summations are then used in an empirical test for

$$A = \frac{|C_1 - X(1,1)|}{|X(1,1)|}, \qquad B = \frac{|R_1 - X(1,1)|}{|X(1,1)|} \qquad (6.16)$$

6.4 Pattern Recognition

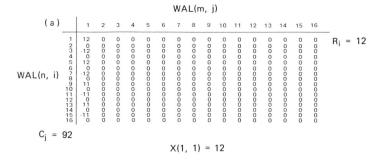

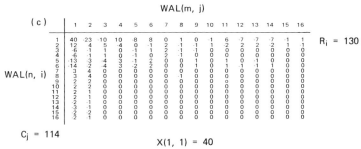

FIG. 6.12. Transformation sub-matrix of (a) a horizontal straightline road feature, (b) a vertical straightline road feature, (c) a road intersection. (Courtesy of the U.S. Army Topographical Laboratories.)

If $A > 0.7$, $B > 0.7$, $C_1 \geq 50$ and $R_1 \geq 50$ (in this particular realisation of the method), then the image may be defined as a road intersection. This may be seen with reference to Fig. 6.12, in which three types of features are compared.

Further summation tests are necessary to determine the angle of straightline roads and intersections to the viewing area and to identify the feature in which it appears in different quadrants of the viewing area. It is also necessary to include ambiguity tests where, for example, the intersection lies at an angle of 45°, which results in very small C_1 and R_1 summations. Diagonal summations are then used as a reference value. A complete series of such tests carried out sequentially and a suitable algorithmic procedure are described by Singleton and Chen [75] and Chen *et al.* [76].

The method is effective for simple linear structures, provided that these do not appear too close to the corners of the viewing area; otherwise, misclassification can occur. The detectable angular resolutions for linear roads and intersections are stated as approximately 11° and 22°, respectively.

6.4.2 Syntactic approach

The syntactic approach is used when the image patterns under consideration are so complex and the number of features needed to describe them so large that the idea of describing the pattern in terms of a hierarchical composition of simpler sub-patterns becomes quite attractive. Examples are found in scene analysis, fingerprint detection and ideograph classification.

An application for which Walsh functions have been found useful is the automatic recognition of written Chinese characters or ideograms. Wang and Shian [77] have pointed out that a mechanised system for Chinese technical writings is now becoming an urgent requirement. China has an ethnic population of nearly 1000 million people producing some 3 billion text words per year, of which less than 1% is currently translated into a European language. The difficult task requires skilled translators, and the basic problem lies in forming a lexicographical ordering of Chinese ideographs *(Kanji characters)* in which there is a need to recognise up to 8000 separate characters [77].

Kanji characters are composed from vertical, horizontal and oblique lines and include features such as corners, T shapes and rectangles, all of which may be classified well through the Walsh or Hadamard transform [78]. In Wang and Shian's implementation, the two-dimensional transformation of the digitised character data is followed by a classifier which puts the patterns into a range of sub-groups. The features of the groups are chosen and compared with a limited set of simple features. A minimum-distance-to-mean classification rule which has the form

$$\sum_{i=1}^{N} (X_i - R_i)^2 \qquad (6.17)$$

is used and is essentially a standard deviation or MSE process. A somewhat similar procedure is carried out by Narasimhan *et al.* [63] for alphanumeric machine print recognition. As with the decision theoretic approach, the difficulty of accurate recognition of shifted or rotated characters has caused the non-invariant forms of the Walsh function to be used, and Wang and Shian [77] and Narasimhan et al. [63] all employ the R transform.

Narasimhan et al. [63] represent each character by a 32 × 32 matrix of digitised data. Only two gray levels, represented by 0 and 1, are allowed for all characters. A two-dimensional R transform of a set of prototype characters is taken through row and column transformation. For each prototype the transform components are arranged in order of decreasing variance and stored in a computer memory. The R transform with its data-compressional capability reduces the identifiable characteristics (features) to a comparatively small number. These values can then be compared in an MSE process with the R transform coefficients through the process shown in (6.17). Identification is realised through the least value obtained from each prototype–signal MSE calculation. Some smoothing of both prototype and signal set has been found to increase recognition accuracy. This differs from Wang and Shian's [77] method in that the prototype values are filtered from noise before recognition and the recognition is invariant to gray-level interchange of the characters (i.e., black/white reversal).

The recognition problem has also been considered by Takahashi and Kishi [79], who use a specialised form of the Hadamard transform for this purpose. In this transform only those coefficients or products of coefficients relating to horizontal, vertical and oblique lines are considered. This enables a compact transform definition of a simplified form of Kanji characters to be obtained (with some loss of orthogonality).

References

1. Andrews, H. C. (1970). "Computer Techniques in Image Processing." Academic Press, New York.
2. Huang, T. S. (1975). "Picture Processing and Digital Filtering." Springer-Verlag, Heidelberg.
3. Simon, J. C., and Rosenfeld, A. (1977). "Digital Image Processing and Analysis." Noordhoff, Leiden, Netherlands.
4. Wintz, P. A. (1972). Transform picture coding. *IEEE Proc.* **60**, 809–819.
5. Ahmed, N., Natarajan, T., and Rao, K. R. (1974). On image processing and a discrete cosine transform. *IEEE Trans. Comput.* **C-23**, 90–93.
6. Connor, D. J., Brainard, R. C., and Limb, J. O. (1972). Intraframe coding for picture transmission. *IEEE Proc.* **60**, 779–800.
7. Knauer, S. C. (1976). Real-time video compression algorithm for Hadamard transform processing. *IEEE Trans. Electromag. Compat.* **EMC-18**, 28–36.
8. Clarke, C. K. P. (1976). Hadamard transformation: Assessment of bit-rate reduction methods. BBC Res. Dept. Report 1976/28, London, England.
9. Pratt, W. K., Chen, W. H., and Welch, L. R. (1974). Slant transform image coding. *IEEE Trans. Commun.* **COM-22**, 1075–1093.

10. Enomoto, H., and Shibata, K. (1971). Orthogonal transform coding system for television signals. *Proc. Symp. Applic. Walsh Functions, Washington, D.C.*, AD727000, pp. 11–17.
11. Lynch, R. T., and Reis, J. J. (1976). Haar transform image coding. *Proc. Nat. Telecom. Conf.*, pp. 44.3-1–44.3-5.
12. Andrews, H. C., and Caspari, K. L. (1970). A generalised technique for spectral analysis. *IEEE Trans. Comput.* **C-19**, 16–25.
13. Narasimha, M. J., and Peterson, A. M. (1978). On the computation of the discrete cosine transform. *IEEE Trans. Commun.* **COM-26**, 934–936.
14. Schaming, W.D. (1974). Digital image transform encoding. RCA Adv. Tech. Lab. Report PE-622, Camden, New Jersey.
15. Haralick, R. M. (1976). A storage efficient way to implement the discrete cosine transform. *IEEE Trans. Comput.* **C-25**, 764–765.
16. Chen, W., Smith, C. H., and Fralick, S. C. (1977). A fast computational algorithm for the discrete cosine transform. *IEEE Trans. Commun.* **COM-25**, 1004–1009.
17. Hein, D., and Ahmed, N. (1978). On a real-time Walsh/Hadamard/cosine transform image processor. *IEEE Trans. Electromag. Compat.* **EMC-20**, 453–457.
18. Ghanbari, M., and Pearson, D. E. (1982). Fast cosine transform implementation for television signals. *IEEE Proc.* **129**, 59–68.
19. Jones, H. W., Hein, D. N., and Knauer, S. C. (1978). The Karhunen–Loève discrete cosine and related transform obtained via the Hadamard transform. *Int. Telemetering Conf., Los Angeles*, pp. 87–98.
20. Srinivassan, R., and Rao, K. R. (1983). An approximation to the discrete cosine transform. *Signal Processing* **5**, 81–85.
21. Jain, A. K. (1976). Some new techniques in image processing. *ONR Symp. Current Prob. Image Sci., Monterey, California.*
22. Yip, P., and Rao, K. R. (1980). On the computation and effectiveness of discrete sine transform. *Comput. Electron. Eng.* **7**, 45–55.
23. Wang, Z. de (1982). Fast algorithm for discrete sine transform implemented by fast cosine transform. *IEEE Trans. Acous. Sound Sig. Proc.* **ASSP-30**, 814–815.
24. Beauchamp, K. G. (1973). "Signal Processing Using Analog and Digital Techniques." Geo. Allen & Unwin, London.
25. Chen, W. H., and Smith, C. H. (1977). Adaptive coding of monochrome and color images. *IEEE Trans. Commun.* **COM-25**, 1285–1292.
26. Ghanbari, M. (1979). Real-time transform coding of broadcast-standard television pictures. Ph.D. thesis, University of Essex, Essex, England.
27. Nicol, R. C., Fenn, B. A., and Turkington, R. D. (1981). Transmission techniques for picture Prestel. *Radio Electron. Eng.* **51**, 514–518.
28. Clarke, K. E. (1980). The application of picture coding techniques to view data. *IEEE Trans. Consumer Electron.* **CE-26**, 568–577.
29. Netravoli, A. N., and Limb, J. O. (1980). Picture coding: A review. *IEEE Proc.* **68**, 366–406.
30. Jalali, A., and Rao, K. R. (1982). A high-speed FDCT processor for real-time processing of NTSC color TV signal. *IEEE Trans. Electromag. Compat.* **EMC-24**, 278–286.
31. Kamangar, F. A., and Rao, K. R. (1981). Interfield hybrid coding of component color television signals. *IEEE Trans. Commun.* **COM-29**, 1740–1753.
32. Heller, J. A. (1974). A real-time Hadamard transform video compression system using frame to frame differences. *Nat. Telecom. Conf., San Diego, California*, pp. 77–82.
33. Jones, H. W. (1977). A conditional replenishment Hadamard video compressor. *Proc. Soc. Photo-Opt. Instrum. Eng.* **119**, 91–98.

34. Jones, H. W. (1976). A real-time adaptive Hadamard transform video compressor. *Proc. Soc. Photo-Opt. Instrum. Eng. Int. Tech. Symp., 20th, San Diego, California,* pp. 2–9.
35. Landau, H. J., and Slepian, D. (1971). Some computer experiments in picture processing for bandwidth reduction. *Bell. Syst. Tech. J.* **50,** 1521–1540.
36. O'Neal, J. B. (1966). Predictive quantising systems (DPCM) for transmission of television signals. *Bell. Syst. Tech. J.* **45,** 689–721.
37. Frieden, B. R. (1975). Image enhancement and restoration, *in* "Picture Processing and Digital Filtering" (T. S. Huang, ed.), pp. 179–246. Springer-Verlag, Berlin.
38. Wang, Z. H. (1982). A simple edge detection by two-dimensional spatial sequency digital filter. *IEEE Proc. Int. Conf. Pattern Recognition, Munich, Germany.*
39. Cappellini, V. Constontinides, A. G., and Emiliani, P. (1978). "Digital Filters and Their Applications." Academic Press, London.
40. Bernabo, M., Cappellini, V., and Fondelli, M. (1980). Edge extraction techniques, *in* "Digital Signal Processing" (V. Cappellini and A. G. Constantinides, eds.), pp. 261–269. Academic Press, London.
41. O'Gorman, F. (1978). Edge detection using Walsh functions. *Artificial Intelligence* **10,** 215–223.
42. Lynch, R. T., and Reis, J. J. (1976). Haar transform image coding. *Proc. Nat. Telecom. Conf., Dallas,* pp. 44.3-1–44.3-5.
43. Sivak, G. (1979). The Haar transform: Its theory and computer implementation. Army Armament Res. Develop. Command NTIS AD-AO70518/6, Washington, D.C.
44. Sivak, G. (1980). Applications of the Haar transform to IR imagery. Army Armament Res. Develop. Command NTIS AD-AO82296/5, Washington, D.C.
45. Shore, J. E. (1973). On the applications of Haar functions. *IEEE Trans. Commun.* **COM-21,** 209–216.
46. Shore, J. E. (1973). A two-dimensional Haar-like transform. Naval Res. Lab. Report 7472. NTIS AD755433, Washington, D.C.
47. Kennedy, J. D. (1971). Walsh function imagery analysis. *Proc. Symp. Applic. Walsh Functions, Washington, D.C.,* AD727000, pp. 7–10.
48. Treitel, S., and Robinson, E. A. (1969). Optimum digital filters for signal-to-noise ratio enhancement. *Geophysical Prospecting* **17,** 248–293.
49. Gubbins, D., Scollar, I., and Wisskirchen, P. (1971). Two-dimensional digital filtering with Haar and Walsh transforms. *Ann. Geophys.* **27,** 85–104.
50. Andrews, H. C. (1970). "Computer Techniques in Image Processing." Academic Press, London.
51. Pratt, W. K. (1972). Fast computational techniques for generalised two-dimensional Wiener filtering. *IEEE Trans. Comput.* **C-21,** 636–641.
52. Hall, E. L., and Kahveci, A. (1971). High resolution image enhancement techniques. *Proc. Two-Dimen. Sig. Proc. Conf., Columbia, Missouri,* pp. 1.5.1–1.5.9.
53. Hazra, L. N. (1977). Real-time image restoration through Walsh filtering. *Optica Acta* **24,** 211–220.
54. Hazra, L. N. (1978). Walsh functions in lens optimisation. *Optica Acta,* **25,** 573–584.
55. Sondhi, M. M. (1972). Image restoration: The removal of spatially invariant degradations. *IEEE Proc.* **60,** 842–853.
56. Smith, E. G. (1978). A new two-dimensional image filtering technique based on a one-dimensional adaptive filtering method. *IEEE Comput. Soc. Conf. Pattern Recognition Image Proc., Chicago,* pp. 113–118.
57. Hunt, B. R. (1975). Digital image processing. *IEEE Proc.* **63,** 693–708.
58. Fu, K. S. (ed.) (1976). "Digital Pattern Recognition." Springer-Verlag, Berlin.
59. Ullman, J. R. (1973). "Pattern Recognition Techniques." Butterworth, London.

60. Harmuth, H. F. (1977). "Sequency Theory: Foundations and Applications." Academic Press, New York.
61. Dinstein, I., and Silberberg, T. (1980). Shape discrimination with Walsh descriptors. *IEEE Proc. Int. Conf. Pattern Recognition, 5th,* pp. 1055–1066.
62. Sethi, I. K., and Sarvarayudu, G. P. R. (1980). Boundary approximation using Walsh series expansion for numerical recognition. *IEEE Proc. Int. Conf. Pattern Recognition, 5th,* pp. 879–881.
63. Narasimhan, M. A., Devarajan, V., and Rao, K. R. (1980). Simulation of alphanumeric machine print recognition. *IEEE Trans. Syst., Man Cybernet.* **SMC-105,** 270–275.
64. Goble, L. G. (1976). Low-pass, spatially filtered Fourier and Walsh transforms as a metric for human pattern perception. *IEEE Proc. Nat. Aerospace Electron. Conf.,* pp. 873–877.
65. Wendling, S., Gagneux, G., and Staman, G. (1976). Use of the Haar transform and its properties in character recognition. *IEEE Proc. Int. Joint Conf. Pattern Recognition,* pp. 844–848.
66. Nemcek, W. F., and Liu, W. C. (1974). Experimental investigation of automatic signature verification. *IEEE Trans. Syst. Man Cybernet.* **SMC-4,** 121–126.
67. Sebestyen, G. (1962). "Decision Making Processes in Pattern Recognition." Macmillan, New York.
68. Reitboeck, H., and Brody, T. P. (1969). A transformation with invariance under cyclic permutation for applications in pattern recognition. *Information Control* **15,** 130–154.
69. Marchesi, C., Giovani, L., and Londucci, L. (1980). A new feature extraction method for ECG ambulatory monitoring based on the Walsh transform. *IEEE Proc. Am. Symp. Comp. Appl. Med. Care, 4th,* pp. 1128–1132.
70. Domokos, L. (1981). Orthogonal transformations for feature extraction in chemical pattern recognition. *Anal. Chim. Acta. Comput. Tech. Optimis.* **5,** 261–270.
71. Meffert, B., Langer, H., and Schubert, D. (1983). Characteristic functions of the sequency technique and examples of the application in information processing. *IEEE Symp. Electromag. Compat., Arlington, Virginia,* pp. 557–562.
72. Chen, P. F., and Seemuller, W. W. (1980). Application of Walsh transforms for topographic feature extraction using a sensory array system. *IEEE Trans. Instrum. Meas.* **IM-29,** 52–57.
73. Chen, P. F., and Seemuller, W. W. (1979). Signal signatures of topographic features using analog technology. U.S. Army Topographic Labo. Report, Dept. ETL-0185, Fort Belvoir, Virginia.
74. Chen, P. F., Rohde, F. W., and Seemuller, W.W. (1979). Prototype image spectrum analyser (PISA) for cartographic feature extraction. U.S. Army Topographic Lab. Report, Dept. ETL-0204, Fort Belvoir, Virginia.
75. Singleton, J. R., and Chen, P.F.(1980). Detecting line-roads and road-intersection patterns at various angles. U.S. Army Topographic Lab. Report, Dept. ETL-0274, Fort Belvoir, Virginia.
76. Chen, P. F., Rohde, F. W., and Seemuller, W. W. (1980). Classification of cartographic features through Walsh transforms. U.S. Army Topographic Lab. Report, Fort Belvoir, Virginia.
77. Wang, P. P., and Shian, R. C. (1973). Machine recognition of printed Chinese characters via transformation algorithms. *Pattern Recognition* **5,** 303–321.
78. Takahashi, K. (1980). Feature extraction method by Hadamard transform. *IEE Proc. Jpn., Tohoku,* pp. 2G–8.
79. Takahashi, K., and Kishi, T. (1980). Feature extraction by specialised Hadamard transform. *IEEE Proc. Int. Conf. Pattern Recognition, 5th,* pp. 1198–1200.

Chapter 7

Communications

7.1 Introduction

A search for increased efficiency in communication methods to meet the expanding requirements for high-volume information transfer has led to an explosion of new ideas and techniques over the past two decades. The application of sequency techniques to communication problems has been particularly rewarding for those digital or two-level design processes in which the characteristics of the Walsh and other orthogonal sequency functions match closely the functional needs. In many cases it has been found that certain characteristics of the orthogonal non-sinusoidal functions permit new ideas in hardware development not previously available.

The impetus for much of the application of these functions to communication can be attributed to the outstanding achievements of Harmuth [1] and his associates. Special reference should be made also to the work of Taki and Hatori [2] and particularly of Hübner [3], who was one of the first to demonstrate some of the practical advantages of the newer techniques in a real working environment. These advantages include efficient multiplexing hardware, reduction of transmission bandwidth and lowering of error rates.

The efficient coding of the transmitted data is also important, and in this the pioneering work of Redinbo [4] in applying sequency methods to communications and control problems should be noted. Finally the contribution of Harmuth [5] and others in opening up the new field of non-sinusoidal wideband communications and radar is one of the major developments in sequency applications during the past decade.

Our discussion of these subjects commences with developments in sequency multiplexing which are central to many communication systems.

7.2 Multiplexing

Multiplexing refers to those techniques that enable simultaneous transmission of many independent signals over a common (i.e., shared) communications channel. We will consider systems of multiplexing based on sets of orthogonal functions used as communication carrier signals. Whilst in theory any set of complete orthogonal functions may be used, practical systems are confined to the use of block pulses and sinusoidal and Walsh functions.

Two well-known systems in common use are *time-divisional multiplexing* (TDM), which uses a set of block pulses as the carrier, and *frequency-division multiplexing* (FDM), which uses a set of sinusoidal waveforms. A third method uses a Walsh function carrier and is known as *sequency-division multiplexing* (SDM). All of these systems are arranged to accept continuously-varying (analog) signals for transmission over the shared channel and to separate these into independent analog signals at the receiving end.

In many modern communication systems the signals to be transmitted are in digital form, and it is convenient to retain this form for the common channel signal as well. Techniques which use a Walsh carrier for these purposes are known as *digital sequency multiplexing* (DSM) systems.

A general multiplexing system which illustrates these techniques is shown in Fig. 7.1, in which a complete orthogonal set of functions, defined over an interval T, is used as the set of signal carriers. This set is designated as $C(i, t)$, where i is the index of the function. The set of input signals x_i corresponds in number with the set of orthogonal functions. To ensure that little variation in signal level occurs during the period T, the signals are low-pass filtered before being applied to the multipliers M.

The products $x_i(t)$, $C(i, t)$ are summed and transmitted over a single

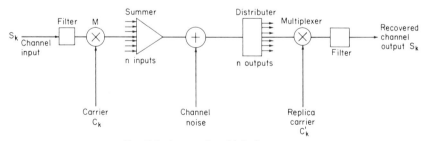

FIG. 7.1. A general multiplexing system.

transmission line to give a composite signal $s(t)$,

$$s(t) = \sum_{i=1}^{n} x_i(t)C(i, t) \qquad (7.1)$$

where $i = 1, 2, \ldots, n$ and n is the number of separate input channels used.

Note that $s(t)$ is an amplitude-modulated signal requiring that the communication channel be linear. Band limiting and phase distortion may impair the process of orthogonality and result in unwanted cross-talk between channels. In addition the communications channel will introduce noise $N(t)$. Consequently the multiplexed signal available at the demultiplexer input is actually

$$\hat{s}(t) = \sum_{i=1}^{n} x_i(t)C(i, t) + N(t) \qquad (7.2)$$

To recover the original analog signals the composite signal $s(t)$ is multiplied by a set of replica carriers $C(k, t)$, where $k = 1, 2, \ldots, n$ — one carrier for each channel output — so that each demultiplexed signal may be represented by

$$x_k(t) = \hat{s}(t)C(k, t)$$

$$= \sum_{i=1}^{n} x_i(t)C(i, t)C(k, t) + N(t)C(k, t) \qquad (7.3)$$

This will produce a large number of products of the n input signals, including harmonic terms, together with a noise signal $N(t)C(k, t)$.

The orthogonality property of the carriers, however, ensures that all these products will reduce to zero for $i \neq k$, so that the output of each demultiplexed channel $k = i$ will be

$$\hat{x}_k(t) = A_k x_i(t) + \hat{N}(t) \qquad (7.4)$$

where A_k is the channel gain and $\hat{N}(t)$ a fraction of the total channel noise.

7.2.1 TDM and FDM

Figure 7.2 shows this principle applied to TDM, where a set of four orthogonal block pulse carriers C_1 to C_4 are modulated by four signal levels $x_i = 1, 2, -1, 0.5$ to give a composite signal $s(t)$ over the sampling time period T. In this multiplex system the channel is shared between the inputs on a time basis. For example, during the interval $(T/4 - T/2)$ the channel will be used exclusively by the $x_i = 2$ input and the other inputs will have negligible effect.

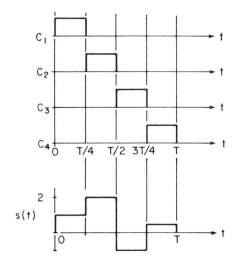

FIG. 7.2. Time-division multiplexing.

Figure 7.3 shows the situation with FDM. Here four sinusoidal waveforms form the carrier signal and the amplitude of these is proportional to the four related input signals $x_i = 1, 2, -1, 0.5$. The composite signals represents the summation of these modulated carriers over the period T. The channel is shared between the inputs on a frequency basis, with each input signal allocated a defined part of the total frequency spectrum.

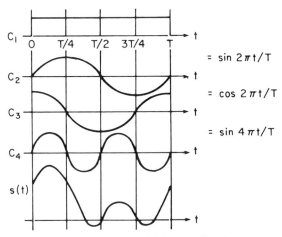

FIG. 7.3. Frequency-division multiplexing.

7.2.2 Sequency-division multiplexing

Apart from the sinusoidal and block functions, there are many other different sets of orthogonal functions which may be used in a multiplex system. Systems using Legendre polynomials [6], Hermite polynomials [7], trigonometric products [8] and Rademacher functions [9] have all been proposed or used. However, the only set of functions which are as efficient as the sinusoidal functions in terms of bandwidth utilisation are the Walsh functions.

Early use of Walsh functions for this purpose simply replaced the sine–cosine carrier signal of an FDM system by a continuous Walsh function signal, that is, a repetition of the pattern shown in Fig. 1.4 at each multiple of the period T, which can be considered as the period of a complex waveform.

The analog Walsh carrier system, shown in Fig. 7.4, is known as a sequency-division multiplexing (SDM) system. A separate function is applied to each of the transmission channels. To avoid distortion due to the limitations in representation of analog signals by discrete samples, the analog samples must first be passed through an aliasing filter (F). Samples are then taken at uniform intervals and constrained by means of a sample-and-hold amplifier (H) to result in a step-shaped signal which can then be transmitted, without distortion, over the sequency multiplex system. To achieve this, the stepped signals are applied to a series of multipliers (M), to which a series of

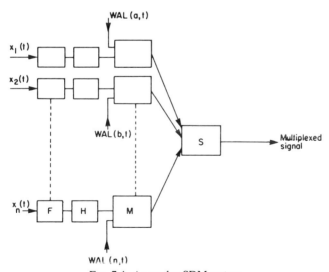

FIG. 7.4. An analog SDM system.

contiguous carrier sequencies are also applied. The output of the multipliers is linearly added before transmission in S.

Demultiplexing using a similar set of synchronised carriers relies on the orthogonal property of the Walsh series and, as with the conventional system using sine–cosine functions, separates out the mixed signals by a process of autocorrelation through multiplication with this replica set of carriers.

One of the advantages of using the Walsh functions as carriers, apart from their ease of generation through digital logic, is that this multiplication process produces only one side-band and not two, as is the case with sinusoidal products. The reason is that the Walsh functions form a group under multiplication so that the product of two Walsh functions is one other Walsh function

$$\text{WAL}(n, t)\, \text{WAL}(m, t) = \text{WAL}(n \oplus m, t) \qquad (7.5)$$

whereas for sine–cosine functions

$$2 \sin \omega_n t \sin \omega_m t = \cos(\omega_n + \omega_m)t + \cos(\omega_n - \omega_m)t \qquad (7.6)$$

This permits side-band filters to be omitted and also simplifies implementation using integrated circuit technology. A number of experimental analog SDM systems have been built and have been described in the literature. A working system designed for 256 voice channels has been described by Hübner [3] of the West German Post Office and other systems by Bagdasarjanz and Loretan [10] and Durst [11].

The main problems with analog multiplexing systems are cross-talk caused by difficulties in achieving accurate synchronisation, frequency bandwidth limitations in the communications channel [3, 12], finite rise time of the Walsh carriers and insufficiently linear analog multipliers.

Synchronisation of the transmitting and receiving systems has been considered by Harmuth [9], who has shown that for numbers of channels that are a power of two the synchronisation problem is simplified if Rademacher functions are used, the orthogonality of which is invariant with a time shift.

Despite the improvements that have been made, other difficulties remain. An almost insuperable one is the widespread development of conventional FDM equipment which causes a justifiable reluctance on the part of communications authorities to change for even quite considerable technical gains. The position is different, however, if we consider Walsh multiplexing of binary signals. The transmission of binary data streams for communication and computer purposes is beginning to impose its own requirements for which the equipment in service is still limited in quantity. This purpose may be achieved through digital sequency multiplexing (DSM).

7.2.3 Digital sequency multiplexing

A number of advantages of DSM over FDM using Walsh functions are noted by Harmuth and Murty [13]. The logic circuits used are small and highly reliable and no setting-up operations are required (e.g., tuning of oscillators and filters). There is also the immunity to burst-type interference arising from the correlation process of demodulation, which is not possible with TDM. Several methods of DSM have been proposed. All of them assign a sequency channel to each input signal, but the form of the multiplexer output varies.

The first method is similar to analog sequency division multiplexing in that the multiplexed signal is multi-level, being the linear combination of Walsh carriers that have been amplitude modulated by the binary digital data streams. This is shown in Fig. 7.5. The data signals to be transmitted x_i to x_n are quantised signals existing over a finite time interval $0 \leq t \leq T$. The carrier signals are Walsh-coded sequences WAL(i, t) to WAL(n, t) having values ± 1 and are mutually orthogonal to each other. As with FDM the pairs of data and carrier signals are multiplied and linearly added. It is assumed that the coded sequences WAL(i, t) to WAL(n, t) forming the replica carriers are stored or recovered at the receiving end and synchronised with the carrier signals. The combined signal is then multiplied separately with each of the replica carriers and the effectively cross-correlated signal may be represented in the case of a particular signal x_k as

$$X_k = \frac{1}{T} \int_0^T x_k \, \text{WAL}(0, t) \, dt + \frac{1}{T} \int_0^T \sum_{i \neq k}^n x_i \, \text{WAL}(i, t) \, \text{WAL}(k, t) \quad (7.7)$$

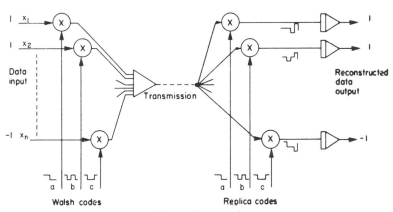

FIG. 7.5. Digital Walsh carrier system.

where the integral sign represents a smoothing operation on the data and channel noise is neglected.

Since the carrier signals are orthogonal the second term reduces to zero, and we can write

$$X_k = \frac{1}{T} \int_0^T x_k \, \text{WAL}(0, t) \, dt \tag{7.8}$$

but $\text{WAL}(0, t) = \pm 1$ so that $X_k = x_k$ and the desired signal is recovered.

A successful system of this type has been installed for fieldtest in Switzerland by Furrer *et al.* [14]. It has applications in various alarm and monitoring systems in which cable connections are employed. The system does, however, suffer from several disadvantages which preclude its wider application. The transmitted signal is a multi-level signal having a variable peak-to-average-power relationship and is therefore susceptible to corruption by noise. The system also requires very linear multipliers, a disadvantage found with the analog system described earlier. These difficulties have led to the development of systems in which the multiplexed transmitted signal is also a binary signal having only two levels. This gives a peak power which is equal to the average power, and hence there is optimum immunity from noise. A further advantage is that it is easier to design simple digital multipliers for multi-level operation to a much higher degree of accuracy than can be achieved with analog multipliers.

Gordon and Barrett [15] have described a digital multiplexing technique of this kind in which only the sign of the multiplexed signal is transmitted. This is possible since with binary signalling it is only necessary to determine the sign of the correlation coefficient for the transmitted information, relative to a single channel, for this to be unambiguously recovered.

Each channel of binary data to be multiplexed is amplitude modulated with one of a set of binary Walsh codes acting as the transmission carrier signals. The summated signal is passed through a hard-limiting device which transmits only the sign of the majority logical value of the modulated channel signal in each time slot. For this reason the system is known as *majority logic multiplexing*. Demultiplexing is carried out by finding the correlation coefficient of the transmitted signal with each of the replica Walsh codes generated at the receiver. The number of channels which may be used with this system depends on the existence of a matrix, the rows of which show a sign-invariant correlation coefficient after being modulated, summed and threshold limited. Such a matrix can be formed from the Walsh functions $\text{WAL}(1, t)$ to $\text{WAL}(7, t)$. Unfortunately, the use of a larger matrix than this does not provide the unambiguous signal transmission obtained with this definition. A way of extending the number of channels through concatenation has also been described by Gordon and Barrett [16]. Here the output of one multiplexer is taken to be the input to the second and so on. Thus each successive demultiplexer is subject to a lower error rate, and itself reduces this rate. An alternative way of increasing the number of channels to a maximum of 49 is suggested by Mukherjee and Mukhopadhyay [17], who used a two-stage transformation process.

With suitably chosen codes for the carriers, the majority multiplexing scheme offers a trade-off between the number of channels in use and automatic error correction. This results from the fact that the magnitude of the correlation coefficient for a given channel increases inversely with the number of channels mixed, so that the correct sign is maintained even in the presence of errors in the multiplexed signal. This is illustrated in Fig. 7.6, which shows the effect of adding an error bit to the multiplexed signal and its elimination during the process of recovery. In this example three out of a possible seven channels are shown. Although some error correction capability can be provided when the system is not fully loaded, it is still necessary for the multiplexer to be able to determine which channels are active. Durst [11] has analysed the allocation of channel power in such a system by plotting the user channel bit error probability against the channel signalling error rate. He has concluded that in a seven-channel system, deterministic errors causing ambiguous performance begin to occur when four or five channels are active.

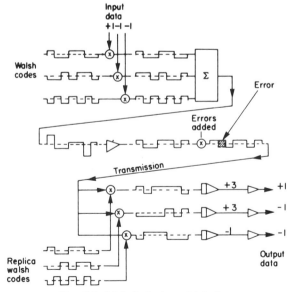

FIG. 7.6. Majority logic multiplexing.

The adaptive majority multiplexer scheme has three advantages, as noted by Schreiber [12].

1. When the system is not fully loaded, the data are redundantly encoded and provide some error-correction capability.
2. The transmitted signal is binary, simplifying the design of transmission equipment.
3. The multiplexer and demultiplexer can be built entirely of digital integrated circuits.

In addition Mukherjee and Mukhopadhyay [17] have observed that, since for each valid code group a one-to-one mapping exists between the original and transformed signal information and furthermore since any change among the rows and columns of the coding matrix maps the original on to an entirely different binary group, the method can also be used in cryptographic data communication. No separate cipher/decipher unit is required, and switching from one mode to another can easily be changed by suitable programming. A number of adaptive majority multiplexers of this type have been constructed for digital telephone signals [18, 19], and an extension to the theory of valid code matrix selection has been made by Changxin [20]. A particularly economic system in terms of bandwidth required is the adaptive Walsh multiplex scheme described by Zheng and Youwei [21] and implemented as a 12-channel digital telephone system. This follows the method described by Jones [22] in which the statistics of the voice signals are used to activate a cut-off circuit so that in the event of a long pause in the transmission of speech signals the channel is removed from the bit-sharing multiplex activity. In Zheng and Youwei's adaptation a method of frame synchronisation continues to operate for the cut-off channels, thus maintaining orthogonality and significantly improving the noise performance of the system without impairing the bandwidth economy achieved.

7.3 Coding

To transmit binary data successfully it is necessary to code them in such a way as to ensure the transmission of a maximum amount of information in a given time with the minimum amount of error. The process inevitably leads to compromise, and the wide variation in coding methods available reflects the gradations of choice between transmission efficiency and levels of intelligibility. Consider first a perfect channel which conveys data unchanged and uncorrupted by extraneous noise, i.e., noiseless coding. The messages to be transmitted are generated by a random variable s of values $s_1, \ldots, s_m$, known as the *source alphabet*. Each s_i consists of symbols $x_1, \ldots, x_n$. These are the code characters taken from a set x known as the *code alphabet*.

In the case of binary transmission x has only two values: 0, 1 or $-1, +1$. Each s_i is assigned a finite set of code characters making up a code word. The collection of code words is called a *code*. For binary transmission this procedure creates a set of binary block codes. The object of noiseless coding is to minimise the average code word length without impairing intelligibility. It is necessary to preserve the uniqueness of a code word in relation to other code words, and some indication must be provided to identify the boundaries between code words transmitted in an uninterrupted stream. Procedures to deal with these and other difficulties are to be found in many standard works [23, 24]. We need to consider here the practical situation of how to achieve reliable transmission through an unreliable channel contaminated with other unwanted signals, i.e., noise, and to see how sequency methods can be employed in this situation. One solution to the problem of noise is simply to repeat the faulty transmission until a correct version is received. This may be effective, provided that we are able to detect the reception of a faulty message and provided that we have a two-way transmission system to ensure repetition upon request. Otherwise we need to *correct* the error as well as to *detect* it. Error detection will be considered first and a summary of some of its salient features given.

A well-known method of detecting single-digit errors is to add another digit, known as a *parity digit,* to each code word. In the method's simplest form the digits of a binary word are inspected before transmission and the added digit is chosen to be 0 or 1 as necessary to make the total number of digits in the 1 state either even or odd, depending on prior agreement. If the word is corrupted in transmission, it is likely that a check on the number of digits received in the 1 state will reveal the existence of single-digit errors. Other more elaborate schemes to detect multiple errors are possible [25].

The use of parity digits is one method of *redundancy coding.* A number of digits are added to a word before transmission and the number and location of digits altered in transmission can be used to permit both error detection and error correction. The value and limitations of such methods in relation to transmission efficiency were first stated by Shannon [26]. He showed that coding methods exist by which information may be transmitted over a noisy channel with an arbitrary small frequency of error, provided that the rate of data transmission is less than a uniquely defined quantity called the *channel capacity C.* Here

$$C = B \log_e(1 + P_S/P_N) \quad \text{bits/s} \tag{7.9}$$

where B is the available bandwidth, P_S the average signal power and P_N the average noise power in bandwidth B.

To achieve nearly error-free transmission at a rate comparable to the channel capacity, very long code words including very many redundant characters are needed. In a practical system we have to compromise with shorter code words and select coding methods to take advantage of known characteristics of the communication channel when it is contaminated with noise.

An example of a useful form of coding for data transmission is the Gray code discussed in Subsection 1.3.4. Here the transition between one code word and the next in an ordered series is always made by changing only one digit. Contrast the change in word format in going from decimal 7 to 8 for the Gray code shown in Table 1.2 and in a pure binary code. The latter requires a change in four digits instead of one, so that an error in any digit could result in a totally different representation. The Gray code is valuable in transmission of slowly varying numerical data for which the expected change can easily be identified with the single-digit changes in the code representation. This is not suitable for the general case of a random variable, however, and redundancy methods are used. Having redundancy in a set of N code words means that some members of this set, a group of K code words, convey useful information whilst the remaining words $(N - K)$ would not appear in noise-less coding and their presence in the received code must indicate that errors are occurring in the communication channel. A useful measure of the effectiveness of a redundant code is due to R. W. Hamming [27] and is known as the *Hamming distance,* which is defined as the number of digit positions by which two members of the K group of words differ from each other. Usually the minimum distance found in the set of K words is meant. This is illustrated by means of the Hamming cube shown in Fig. 7.7. The cube represents all possible codes from a set of $N = 8$ code words. The corners of the cube represent the eight possible code words, and movement along one edge from one code word to the next corresponds to a change of one digit. If we consider an even parity code consisting of four code words 00, 01, 10, 11 to which are added parity bits, we have 000, 011, 101, 110 as the set of K valid codes. The remaining $(N - K)$ words 001, 010, 100, 111 represent invalid codes only occurring with transmission error. In Fig. 7.7 the K codes are situated at opposite corners of a cube face and are separated by a corner representing one of the $N - K$ invalid codes. The Hamming distance between valid codes is 2 in this case. Whilst the Hamming distance is widely quoted as a *good* parameter for a coding system, a large Hamming distance does not necessarily mean a low probability of error since this also depends on the energy of each transmitted character [1].

7.3.1 Error control

As we have seen, error detection, particularly of single isolated errors, is relatively simple. A parity check may be adequate for not-too-critical trans-

7.3 Coding

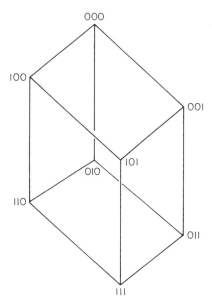

FIG. 7.7. The Hamming cube.

missions for which a repeat facility is available. A more complex example is an m-out-of-n code, where m represents the number of bits used for data and $m - n$ the bits reserved for errors in a given word of n bits. In one particular representation $m = 4$ and $n = 8$ and each valid word must have four bits at the 1 state and four at the 0 state. The number of ways of choosing m from n is

$$n!/m!(n - m)! \qquad (7.10)$$

so that there would be 70 valid states out of a possible 256. If the errors were uniformly distributed, this code would be about 1.8 times as effective as a 7-bit plus parity code having 128 valid states out of 256.

By adding still more redundancy it is possible to correct as well as to detect isolated errors. Single-error detection is feasible for a Hamming distance of 2. If this is increased to 3, then single errors can be detected *and* corrected. A minimum Hamming distance of 4 permits the detection of two errors and correction of one. A distance of 5 permits the correction of two errors, and so on. Where extensive correction of errors is needed, the codes begin to get very long and channel capacity is seriously eroded.

A considerable improvement in performance is obtained if the *group codes* originally proposed by Elias [28] are used. These can be used to transmit information at a rate close to the channel capacity with an arbitrarily small probability of error. The encoding scheme is relatively simple and decoding is through a maximum likelihood detector which although complex is well suited to computer operation.

A feature of the group codes is that a code alphabet for the group may be defined by a matrix relating check digits with valid information digits, so that matrix calculation methods can be used to determine the most likely of a range of corrupted data possibilities to a given valid code word. Error correction is therefore a question of probabilities and the likelihood in statistical terms of replacing a faulty code word with its uncorrupted original form.

7.3.2 Reed–Muller codes

An important class of group codes are the Reed–Muller codes, which are closely related to the form of the Walsh or Hadamard matrix [29]. For code words of length $n = 2^p$ it can be shown that $(2^{p-r-1} - 1)$ errors can be corrected in a reasonably straightforward manner by majority testing of redundant relations [30]. Here r is related to the size of the code where the number of code words is 2^k and

$$k = \sum_{i=0}^{r} p_i, \quad r < p \tag{7.11}$$

If $r = 1$ then the set of code words is said to be a Reed–Muller (R-M) alphabet of the first order. The number of check digits is $p - k$ and the minimum Hamming distance is

$$d = 2^{p-r} \tag{7.12}$$

Let us consider a first-order R-M alphabet in which the word length is given as $n = 2^3 = 8$. It follows from Eq. (7.11) that $k = 1 + 3 = 4$, so that the alphabet will contain $2^4 = 16$ words each of 8 digits. This R-M alphabet is denoted as a (8, 4) alphabet and is illustrated in Fig. 7.8. The minimum Hamming distance is $2^2 = 4$, which will allow the correction of $2^{3-1-1} - 1 = 1$ error. If Fig. 7.8 is now compared with Eq. (1.47), it will be seen that the R-M alphabet can be constructed very easily from the Walsh matrix since the first half of the code words are the functions WAL(s, t) with $s = 0, 1, \ldots, 2^{k-1}$ and the second half of the functions are $-$WAL(s, t). We thus see that the Reed–Muller alphabets belong to the class of orthogonal alphabets.

Several decoding methods for orthogonal codes are described [31–33] in which the orthogonal relationships in the code structure permit fast transformation procedures to be used.

7.3.3 Variable-word-length coding

Considerable interest has been shown in recent years in adapting the code used to the statistical properties of the transmitted signal. It has been found,

	S	0	1	2	3	4	5	6	7
	1	+1	+1	+1	+1	+1	+1	+1	+1
	2	+1	+1	+1	+1	−1	−1	−1	−1
	3	+1	+1	−1	−1	−1	−1	+1	+1
	4	+1	+1	−1	−1	+1	+1	−1	−1
	5	+1	−1	−1	+1	+1	−1	−1	+1
	6	+1	−1	−1	+1	−1	+1	+1	−1
	7	+1	−1	+1	−1	−1	+1	−1	+1
R-M codes	8	+1	−1	+1	−1	+1	−1	+1	−1
	9	−1	+1	−1	+1	−1	+1	−1	+1
	10	−1	+1	−1	+1	+1	−1	+1	−1
	11	−1	+1	+1	−1	+1	−1	−1	+1
	12	−1	+1	+1	−1	−1	+1	+1	−1
	13	−1	−1	+1	+1	−1	−1	+1	+1
	14	−1	−1	+1	+1	+1	+1	−1	−1
	15	−1	−1	−1	−1	+1	+1	+1	+1
	16	−1	−1	−1	−1	−1	−1	−1	−1

FIG. 7.8. A (8, 4) Reed–Muller code.

for instance, that in order to obtain the best performance for a given transmission it is necessary to represent some data samples more accurately than others. This implies a *variable word length* for the samples transmitted. By allocating word lengths according to the expected variation in the amplitude of the Walsh-transformed coefficients, it has been found that reduced quantisation error is realised compared with a PCM system having constant word length equal to the average word length of the transform coding system [34].

A further development is described by Redinbo [35], who considers the possibility of using Walsh spectral analysis of the state of the communication channel to permit the development of an automatic coding system. In the spectral analysis system, Redinbo describes a system of minimum mean-square error codes by using the results of Walsh analysis to optimise the operation of the binary error-detecting codes used. These codes are based directly on the statistics of the communication channel which is continually analysed by the use of the Walsh transform. Since the Walsh transforms of the channel statistics determine the coding rules, the equipment for real-time processing to determine these rules need not be complex. For example, a flat sequency spectrum is one indication of the optimisation of such channels. It is suggested that these adaptive methods could enable sustained optimum or near-optimum transmission to be achieved, even on channels in which the statistics are slowly varying.

In a later paper by Redinbo and Cheung [36] a system design for an unequal error correction system is described. Usually in channel coding systems when the channel noise temporarily exceeds the error-correcting capabilities of the code, the result is unpredictable and serious degradation can occur. If, however, a system is used which affords an unequal error protection to the information digits, the overall system performance can be made to degrade in a statistically gradual manner.

The basic design technique is to transform the data by using the FWT and in the decoding process to carry out error correction with a code containing n data values by numerically combining the output from $n + 1$ finite impulse response filters. By making the filter weights relate to the measured channel response, it is possible to modify the combinations in such a way as to minimise the MSE between input and output digits. This has the effect of guaranteeing various levels of correcting ability for the more significant digits during temporary overload conditions.

Transformation coding has also been used to reduce the required channel bandwidth and hence increase channel capacity (Eq. (7.9)). As a consequence of Parseval's relationship giving equality between signal energy in the time and sequency domains, it is quite feasible to transmit Walsh-coded signals in the sequency domain in place of digitally-coded samples of the signals. Since the Walsh series converges quite rapidly, it is possible to transmit fewer coefficients without sacrificing much signal energy content and a reduction in transmitted bandwidth is obtained [37].

7.4 Non-sinusoidal electromagnetic radiation

For more than a decade there has been considerable interest in the radiation and detection of *non-sinusoidal waveforms*. This is because of the difficulties that have been encountered with conventional sinusoidal carriers as the transmission frequencies extend into the gigahertz region and because of the increasing use of digital transmission methods which favour the adoption of bi-level techniques in their implementation. The transmission problems are most clearly recognised in the case of radar applications of all types, and it is here that the main thrust of development and application has taken place.

A major difference between the use of sinusoidal and non-sinusoidal carriers is, of course, the bandwidth occupied by the transmission, but we need to look at this in terms of a *relative bandwidth* and ask how the total bandwidth is utilised to transmit information.

The relative bandwidth is usually defined as a ratio of bandwidth to carrier frequency, but since some of the applications we will be discussing do not actually make use of a carrier the more general definition given by Harmuth [5] will be used:

$$\eta = (f_h - f_l)/(f_h + f_l) \qquad (7.13)$$

where f_h is the highest and f_l the lowest frequency of interest.

Typical radio signals used for communications or radar have relative bandwidths of the order of $\eta = 0.01$ or less. In using sinusoidal methods of communication this is a necessary requirement since the phenomenon of resonance disappears with much larger values of η and with it the amplification and frequency selection afforded by the tuned circuit.

Most of the signals we desire to transmit, e.g., sound and vision signals, have η values close or equal to 1.0. A typical solution for transmission purposes is to transform these large relative bandwidths into smaller relative bandwidths by the use of a modulated sinusoidal carrier. This translation of the bandwidth occupied by the signal into a much higher region has been accepted for economic and practical reasons but is now becoming less realistic with the much wider bandwidths demanded by high-resolution radar and in areas dominated by high communication channel density.

A natural limitation is placed on this process of increasing the carrier frequency to achieve a small relative bandwidth by the attenuation effects of rain, fog and molecular absorption when the carrier frequency begins to exceed 10 GHz [38]. Thus if radar transmissions are limited to this maximum carrier frequency, then the resolution is also limited to about 1% of that possible if this bandwidth is utilised directly with η values closer to unity [32]. Similarly for communication it is entirely feasible to increase the channel capacity of a communication system by adopting higher values of η if an appropriate technology is applied [39].

The two techniques that have emerged to attain these results are known as *impulse radar* and *spread-spectrum communication*. Both have their origins in the application of sequency techniques through the use of non-sinusoidal functions, originally block and Walsh functions.

The theoretical work supporting a study of non-sinusoidal electromagnetic radiation has been established by Harmuth [32], who also made early contributions to the design of experimental transmission and reception equipment (Lally *et al.* [40]). A transmitter for electromagnetic Walsh radiation is also described by Fralick [41], who used horn aerials for both transmission and reception. The more difficult problems of the reception of Walsh waves are discussed by Frank [42], who follows the earlier work of Harmuth. The current position for this preliminary work was summarised in a paper by Harmuth [43] in which he noted several basic differences between sinusoidal and non-sinusoidal electromagnetic radiation which could be used. These are the following:

1. The technology for implementation is different. Pure sinusoidal waveforms are relatively difficult to generate. Non-sinusoidal waveforms require only suitable switch matrices operating in the nanosecond region currently achievable with solid-state devices.

2. The differentiation of a sinusoidal function yields a shifted sine func-

tion (actually a cosine function) for the same frequency, whilst the differentiation of a Walsh function yields a differently shaped function which can be recognised uniquely.

3. The summation of sinusoidal functions having arbitrary amplitudes and phases but equal frequency yields a sinusoidal function of the same frequency. Walsh functions are summed differently, so that interference effects do not behave in the same way.

4. The Doppler effect can transform a sinusoidal function into another for any velocity ratio v/c. With a Walsh function a threshold ratio of $|v/c| \geqslant 3.5$ is necessary before a transformation occurs to another Walsh function of the same system.

5. Sinusoidal waves exhibit polarity symmetry; that is, a reversal in amplitude produces the same effect as a time shift. Walsh waves do not have polarity symmetry.

Since the early 1970s many of these differences have been exploited for non-sinusoidal communication and radar. New techniques have been developed and extensions made to sequency theory applied to non-sinusoidal waveforms.

The most significant of these are the new designs of selective receivers capable of detecting sequency waveforms; a study of the effects of non-sinusoidal transmissions on conventional transmissions using the same frequency band; and techniques for impulse radar, particularly into-the-ground radar, which has developed into a new and valuable exploratory tool. The problems of radiation and design for single- and multiple-aerial systems have also received considerable attention. These various developments will be reviewed next, commencing with the facilities required for spread-spectrum communication.

7.4.1 Spread-spectrum communication

Two methods have been considered for communication with non-sinusoidal waveforms. The first is to modulate these waveforms with a signal which may consist of a group of signals sharing a band extending over 1 GHz, i.e., spread-spectrum communication [44]. The second is to transmit directly the non-sinusoidal waveform, which itself contains the information to be transmitted [5].

In the spread-spectrum communication system advocated by Harmuth [45] the carrier is a non-sinusoidal waveform consisting of either a set of block pulses or orthogonal Walsh functions. The latter would be preferred if a set of non-interfering signals are to be transmitted sharing the same frequency spectrum. The transmitted information is carried by amplitude

7.4 Non-Sinusoidal Electromagnetic Radiation

modulation of the transmitted rectangular pulses. If a set of orthogonal Walsh functions are used, and provided that proper synchronisation is carried out, the desired function can be distinguished (by using a correlation process of finite duration) from an accompanying number of undesirable functions from the same set. A second advantage of orthogonal Walsh functions compared with block functions is that for digital sampling, which is desirable in some sections of the receiving process, it is possible to work closer to the Shannon limit, thus improving the efficiency of the system [39].

Walsh waveform detection can comprise four stages, each of which can be identified as a separate electronic process (Fig. 7.9). The remainder of this subsection is a summary of the general system developed by Harmuth.

The first stage consists of a radio sequency filter and amplifier (RSF). The principle is illustrated in Fig. 7.10. A periodic signal of amplitude A arrives at the input a of a summing amplifier. The signal is transmitted around a feedback loop having a delay which corresponds to the period T of the desired signal. The delayed pulse will thus arrive at the amplifier input b in time to add to the second period of the signal, and a signal with amplitude $2A$ will proceed to circulate around the loop. Thus signals of the correct period T will add to produce an amplitude nA after n periods have been received, whilst noise and signals of the incorrect period will build up more slowly. In a practical arrangement the delay is constructed of non-dispersive coaxial cable and several serial resonant loops linked by means of hybrid couplers. By adjusting the delay periods in each resonant loop, the RSF stage can be made to accept and amplify signals of a band centered around the period T [32, 43, 45].

The amplified signal consisting of a modulated Walsh carrier needs to be subject to a discrimination process to distinguish between signals having the same period T but different time variation during the period. It is convenient to carry this out digitally [45] so that a sequency convert (SC) or pulse-stretching stage is inserted before the waveform discriminator (WD). The sequency converter converts the period T into a much longer period $2^n T$. This allows sampling of the amplified waveform to be carried out more easily. A second filter–amplifier similar in principle to the RSF but operat-

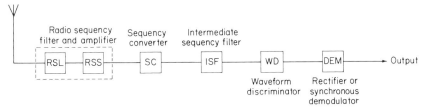

FIG. 7.9. Stages of a receiver for non-sinusoidal waves.

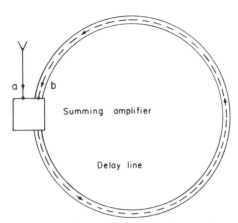

FIG. 7.10. Selective extraction of a periodic waveform from noise or periodic waves with a different period.

ing at the longer period 2^nT can also be included in the design preceding the waveform discriminator.

Waveform discrimination is carried out through a process of correlation with a local carrier of the same sequency. Since the Walsh carrier contains several zero crossings in a given time period, it is essential to select the correct phase for this correlation procedure. In the gated analog circuit described by Frank [42] a search procedure is carried out to ensure correct synchronisation. In Harmuth's digital circuit the waveform discriminator acts essentially as a sliding correlator and is limited to discriminate fairly low-order sequency functions.

Finally the discriminated waveform is demodulated by either a rectifier or a synchronous demodulator, as in the case of a sinusoidal receiver, and the original signal recovered.

The use of spread-spectrum communication in which a high relative bandwidth is employed and shared frequency bands used has opened the possibility of providing additional radio channels for radio and television broadcasting without replacing existing equipment. This implies that the wide-band transmission of the spread-spectrum service will occupy some or all of the spectra of existing sinusoidal carrier systems and degrade its signal-to-noise ratio [46]. Some degradation of existing radio transmissions will inevitably occur in the form of an increased background noise but can be shown to be small and in some cases negligible. This is because the power of the spread-spectrum transmission will be distributed fairly equally over a very wide frequency. In the case of several simultaneous spread-spectrum transmissions these will appear to the receiver for sinusoidal waves like thermal noise. Provided that these transmissions are not too numerous, it

7.4 Non-Sinusoidal Electromagnetic Radiation

appears possible, at least theoretically, to restore the original signal-to-noise ratio by an increase in transmitted power of a few decibels. [39].

7.4.2 Impulse radar

Early attempts to apply sequency methods to radar systems considered the consequences of replacing a sinusoidal wavetrain, modulated or unmodulated, by a Walsh carrier. Lackey [47] observed that the resolution of point targets can be enhanced if the target area is illuminated with a Walsh wave rather than a sinusoidal signal. The effect of a secondary target, close to the required point target, may be seen during the entire period of the transmitted Walsh signal, which is not the case for a single period of a sinusoidal signal. This is shown in Fig. 7.11. Here (a) and (b) show the signals reflected from two target areas T_1 and T_2, where a sinusoidal radar-pulse train is used. The summation signal (c), which is the signal observed by the radar receiver, shows very little sign of the existence of a second target, except for a slight discontinuity of the beginning and end of the pulse train. On the other hand, a Walsh pulse waveform reflected by the two targets as in (d) and (e) will sum to give (f). The difference between a reflection from two separate targets in terms of their summation is no longer a small perturbation of the reflected signal but a major change in the summed and reflected waveform and points to an increase in resolution for the Walsh case. The received patterns of an amplitude-modulated radar transmission for sinusoidal and Walsh carriers were also considered by Harmuth [48]. These transmissions consist of a block pulse modulation of the carrier permitting a short train of sinusoids or pulses to be radiated. Thus, in the case of a sinusoidal carrier, the radiated signal has the form

$$\cos 2\pi t/T \quad \text{between} \quad -mT/2 \leq t \leq mT/2 \quad (7.14)$$

and for the Walsh carrier

$$d\,\text{WAL}(2, t/T)/dt \quad \text{between} \quad -mT/2 \leq t \leq mT/2 \quad (7.15)$$

where T is the period of the carrier and m the number of cycles contained within the period of the block pulse modulation.

In order to detect the time differences (and hence distance travelled) between the transmitted and reflected signals, the autocorrelation of the received amplitude-modulated carrier is used. Autocorrelations are shown in Fig. 7.12 for sinusoidal and Walsh carriers. The important difference between the two functions is that in the Walsh case fairly long sections of zero value are found, whereas in the sinusoidal case a continuous curve is produced. This is significant in the case shown in Fig. 7.11, where two reflectors

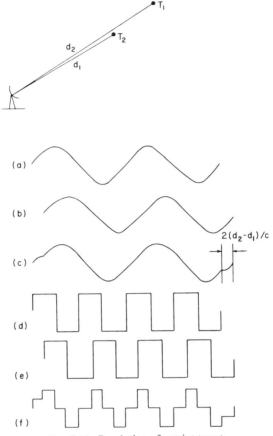

FIG. 7.11. Resolution of a point target.

are considered. The propagation time difference needed for detection differs in the two cases. In using a sinusoidal carrier, a propagation time of about $t_p = mT$ is required, and in fact there is little advantage in using the full-line signal, so that envelope detection (shown dotted in Fig. 7.12) may be used. The Walsh carrier allows discrimination with a propagation time difference between the intervals $T \leq t_p \leq T/2 - \Delta t$, $T/2 + \Delta t \leq t_p \leq T - \Delta t$, etc. If Δt is short compared with T, then a considerable theoretical improvement in range resolution can be obtained.

Some work on the resolution performance of a radar system using Walsh modulation was also carried out by Griffiths and Jacobson [49]. This is described in terms of the *radar ambiguity function* proposed by Woodward [50], which gives a measure of the suitability of a given modulation waveform for radar purposes. These researchers show that the ambiguity infor-

7.4 Non-Sinusoidal Electromagnetic Radiation

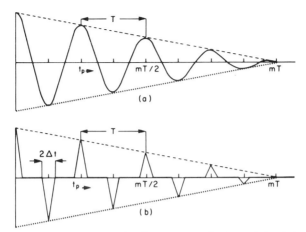

FIG. 7.12. Normalised autocorrelation for (a) an amplitude-modulated sinusoidal carrier and (b) an amplitude-modulated Walsh carrier.

mation for all $N = 2^p$ Walsh functions is contained in a knowledge of the ambiguity for p Walsh basis functions. Therefore, a knowledge of relatively few ambiguity relationships for a Walsh basis series enables a wide range of radar modulation waveforms to be developed that have specific ambiguity properties. The ambiguity functions obtained for the higher-order Walsh functions are surprisingly good and rival the performance of pseudo-random Barker codes, often used for this purpose.

Other early approaches to the utilisation of Walsh waves in radar signal generation include synthesisation methods in which a number of Walsh functions are added to provide a modulating waveform having otpimum characteristics. Some analytical work has been carried out by Rihaczek [51] on these optimum waveforms. The choice of the optimum radar waveform is found to be greatly dependent on the exact nature of the target and type of discrimination required. The formulation of the modulating waveform required demands a generation process which is flexible and easy to mechanise. For reasons given earlier, the Walsh sequence has particular advantages in this respect. Fast digital Walsh transformation techniques for this purpose are discussed by Griffiths and Jacobson [49].

Whilst most of these early investigations looked into the possibilities of using complete Walsh or Rademacher series, the more recent developments have considered instead the transmission of very narrow pulses by utilising a frequency band extending from 1 to ~ 10 GHz. This has become known as *impulse radar* or *carrier-free radar*, and its main characteristic is the use of non-sinusoidal signals having large relative bandwidth.

One extremely successful application is in *into-the-ground impulse radar*, which was first proposed by Cook [53] in 1960. Inhomogeneities in the ground make the attenuation of electromagnetic waves increase with frequency very rapidly, just as raindrops and fog droplets in air inhibit high-frequency radio transmission. In order to provide an acceptable resolution, such into-the-ground radars must use pulses with a duration of about 1/ns, and for reasons given earlier this is not a practical proposition with carrier-borne techniques.

Figure 7.13 illustrates the operation of this type of radar. The radar is mounted such that it can traverse the surface of the ground. A pulse is radiated into the ground at intervals of the order of microseconds to milliseconds. A layer at the depth D with a discontinuity of the dielectric constant ϵ, magnetic permeability μ or the conductivity σ will reflect the signal. The range of detection extends over $2D \tan \beta$ at level D, where the beam angle of the radar equals 2β and the discontinuity will be seen as a band of width $[D/\cos (\beta - D)]$ at the depth D. A number of examples of actual recordings are given by Harmuth [5]. Note that this technique will detect not only buried metallic objects but also other devices such as pipes made of concrete, clay or plastic which exhibit a change in ϵ, μ or σ from the surrounding material. The first commercial application of this principle was described by Chapman [54], and many other papers have been published since his giving details of measurements taken in a variety of environmental conditions (see the references in Harmuth [5, p. 32]).

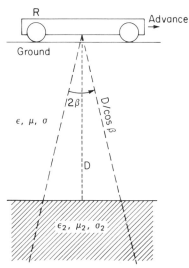

FIG. 7.13. Probing of an extensive layer having discontinuity of ϵ, μ or σ. (From Harmuth [5].)

7.4.3 Radiators

A difficult problem in the development of radar and radio communication equipment using large-relative-bandwidth techniques is the design of an effective aerial system. Existing theory and design are based mainly on resonant dipoles and hence are more appropriate to carrier-borne small-relative-bandwidth systems. Some attempts have been made to use wide-bandwidth methods such as biconical, horn and log-spiral aerials, often fairly successfully, although the transmission efficiency is low [55, 56]. Improved results have been obtained by new techniques not based on Fourier representation of a pulse-shaped waveform. The most promising of these are the *current-loop dipole* and the *travelling-wave aerial* [32, 57, 58].

Hertzian electric and magnetic dipoles can both radiate non-sinusoidal waveforms. The electric dipole (Fig. 7.14a) requires a very high driving voltage to obtain a large current $i(t)$ in the transmitting element. It has been shown elsewhere that high $i(t)$, or rather its derivative di/dt, is necessary to obtain a high radiated field strength [57]. Magnetic dipoles (Fig. 7.14b) achieve large energising currents with low applied potentials, and these are better suited to semiconductor technology. Many practical magnetic dipoles and arrays of dipoles have been used in developed radar systems, and the theory of their design is now well understood [32].

The average radiated power of an array of n interacting dipoles increases proportionately to n^2, whilst the average power radiated per dipole increases proportionately to n. Hence it is usual to find arrays of magnetic dipoles used for non-sinusoidal radiators. Generators for Walsh or pulse-shaped currents are very inexpensive since they consist of semiconductor switches feeding positive or negative currents into the dipole, and so the provision of multiple driving circuits does not present a problem. There are difficulties, however, in supplying these currents in such a way as to avoid radiation from the associated feeders. A way of overcoming this is shown in the arrangement

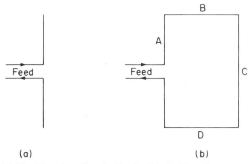

FIG. 7.14. (a) The Hertzian electric dipole, (b) the Hertzian magnetic dipole.

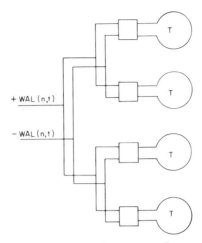

FIG. 7.15. Feeder arrangements for an array of magnetic dipoles.

due to Lally *et al.* [40] given in Fig. 7.15. Four Hertzian magnetic dipoles are energised through switching transistors to which the driving voltages WAL(n, t) and $-$WAL(n, t) are applied (a negative value of WAL(n, t) reverses the current flow). By including a separate power transistor or switching dipole within each loop, the radiating power is confined to the loops, thus minimising feeder-line losses through radiation. Two- or three-dimensional arrays follow the same principle. Multidimensional arrays simulate a point source and are, therefore, suitable for location at the focal point of a parabolic reflector. Stacked magnetic dipoles of this type are not suitable for very short switching times because of transmission delay effects around the loops, although the interlaced feeding mechanism has been adopted in most later developments.

The current loop, or magnetic dipole, shown in Fig. 7.14b has one quite obvious disadvantage which drastically reduces its effectiveness as a radiator in the far field. This is the cancelling effect of contrary direction of currents flowing in two parallel wires A and C. In a similar design used for sinusoidal waves, this is known as a *folded dipole*. Since the radiator is a resonant one, the reversed direction of currents in A and C is compensated by reversal in polarity of the current in the positive and negative halves of a sinusoidal current. This cannot be done for non-symmetric pulse currents, and since the currents in A and C flow in opposite directions, we have not dipole radiation but quadripole radiation, which typically carries much less power [32]. To overcome this a conducting sheet of metal can be inserted between A and C, as shown in Fig. 7.16. To minimise the parasitic reradiation from this sheet derived from dipole C, an absorbing material such as ferrite (shown as F in the diagram) can be placed ahead of the sheet [59].

7.4 Non-Sinusoidal Electromagnetic Radiation

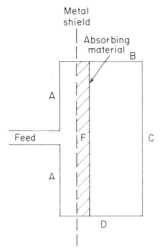

FIG. 7.16. Conversion of a magnetic loop radiator into an electric dipole radiator.

A different principle for obtaining dipole radiation from a current flowing in a loop is to vary the dimensions of the two radiating sides of the loop. The forward side C is constructed of a wide metal sheet, whilst side A is made of thin wire. The effect of this design is that radiation from dipole A is shielded in the forward sector by the width of the metal sheet dipole C.

If it is possible to use polarity-symmetric currents such as those appropriate to Rademacher waveforms, then, as in the folded dipole used for sinusoidal waveforms, the current direction is effectively reversed for the two opposite sides of the loop and a high-power radiating array is obtained. This is shown in Fig. 7.17, which has been applied as a compact radiator for the into-the-ground radar application described earlier [54]. Here a number of magnetic loops are driven in unison by an input digital signal. The dipoles A

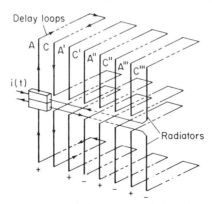

FIG. 7.17. High-power radiating structure for into-the-ground radar.

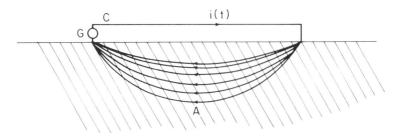

Fig. 7.18. Long-wave aerial with earth return.

and C of the loops are brought close together by folding the loop, and the connecting lines B and D are arranged to give a delay equal to $T/2$, where T is the period of pulse separation. Thus the phase reversal of the applied current pulse equals that of the current reversal around the loops so that the radiation from A, C, A', C', etc., will be in phase. The positive and negative signs below the dipoles indicate that a Rademacher wave WAL(7, t) will be radiated. One may readily see that this radiating structure can be made to radiate a variety of Walsh waves by changing the interconnection of the dipoles via the delay loops.

Other methods that have been used for non-sinusoidal radiation are assemblies of helical radiators in which the direction of helix rotation is made to correspond with the $+1$ or -1 of the energising Walsh function [32] and the use of long-wire radiators [32, 58]. This latter type of aerial (the travelling-wave aerial) is widely used for the radiation of long waves and is effectively aperiodic for microwave pulses. This is shown in Fig. 7.18. The dipole C is a long wire along which the transmitted current $i(t)$ is propagated. The return path via the earth constitutes the dipole A. Since the physical structures of the two paths are so different the dipole-mode radiation is not completely cancelled and an effective broadside radiation occurs [60].

7.4.4 Radiator arrays

The assembly of a number of non-sinusoidal radiators to form an array provides advantages other than a simple increase in radiated efficiency. As with arrays used with sinusoidal (carrier) radiation, directional properties for the radiated or detected signal are inherent in the design. These properties differ for the non-sinusoidal and sinusoidal case, however. The reason for this lies with the addition properties for the two kinds of signal discussed in Chapter 2. Whereas with the sinusoidal functions the sum of two or more sine functions with the same but arbitrary phase and amplitude yields a sine function of the same frequency, the addition of non-sinusoidal functions

7.4 Non-Sinusoidal Electromagnetic Radiation

gives rise to a modulo-2 addition of the functions which can change the shape of the resulting function quite considerably. This has been demonstrated by Hussian [61] in respect to the sinusoidal and non-sinusoidal functions, and he has described some interesting advantages for the latter functions.

Sinusoidal array transmission for a reasonable number of elements gives rise to a beam pattern in terms of angular incidence θ in broadside direction as

$$V(0, \theta) = [\sin(\pi L\theta/\lambda)]/(\pi L\theta/\lambda) \qquad (7.16)$$

where L is the array length, λ is the wavelength of the received wave and the resolution angle for the beam is

$$\epsilon = kcT/L = kc/fL \qquad (7.17)$$

where T is the period, f the frequency, c the phase velocity of the sinusoidal wave and k an arbitrary constant usually chosen to be 1. The corresponding resolution angle for distortionless rectangular pulses is derived by Hussain [61] as

$$\epsilon = kc/\Delta fL \sqrt{P_S/P_N} \qquad (7.18)$$

where Δf is the nominal bandwidth of the pulse (taken as the reciprocal of the pulse width $1/\Delta T$) and P_S/P_N is the signal-to-noise power ratio.

The interesting feature of this derivation is that it indicates that the resolution for non-sinusoidal driven arrays is dependent on the pulse power as well as array length and signal bandwidth. Hence one can obtain a small resolution by increasing the signal power in contrast to sinusoidal signals where an increase in signal power brings no improvement in resolution angle.

The most significant difference between the two kinds of functions is, however, the behaviour of the transmitted signal *off the beam axis*. This is shown for non-sinusoidal array transmission in the wave zone in Fig. 7.19. The radiated pulses will be shifted in time relative to each other and summed to give a received pulse shape which varies with the radiation angle ϕ. On the array axis on which the radiation angle is zero, no time shifts are involved and a rectangular pulse is obtained. For increasing values of radiation angle, the rectangular pulse changes via a trapezoidal pulse to a triangular shape and a series of trapezoidal waveforms of lower peak amplitude and longer duration [32].

The slope of the ramps of the trapezoidal and triangular pulses relates quite accurately to angle of transmission, and this fact has been used by Hussain [62] to determine angular resolution through a correlation process with a stored table of calculated values. This procedure permits the development of scanning radar techniques which do not rely on mechanical or electronic beam-steering methods to search for a target.

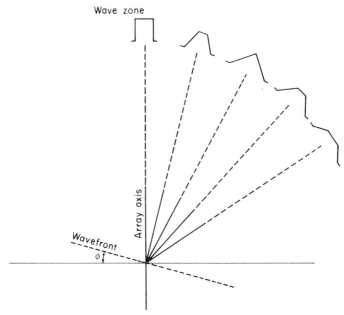

FIG. 7.19. Time variation of pulses in the far zone of a non-sinusoidal array transmission. (Based on Hussain [61].)

For accurate summation of reflections from a radar target using carrier-free pulses it is necessary to consider also reflections from a target which lie off the beam axis and will produce the trapezoidal or triangular reflections shown in Fig. 7.19. Signal processing methods involving a linear regression technique or a correlation technique to carry out this beam-forming process are described by Hussain [62].

References

1. Harmuth, H. F. (1972). "Transmission of Information by Orthogonal Functions," 2nd ed. Springer-Verlag, Berlin.
2. Taki, Y., and Hatori, M. (1966). P.C.M. communication system using Hadamard transformation. *Electron. Commun. Jpn.* **49**, 247–267.
3. Hübner, H. (1971). Analog and digital multiplexing by means of Walsh functions. *Proc. Symp. Applic. Walsh Functions, Washington, D.C.*, AD727000, pp. 180–191.
4. Redinbo, G. R. (1972). Linear mean-square error codes. *Proc. Symp. Applic. Walsh Functions, Washington, D.C.*, AD744650, pp. 330–336.
5. Harmuth, H. F. (1981). "Nonsinusoidal Waves for Radar and Radio Communication." Academic Press, New York.
6. Ballard, A. H. (1962). A new multiplex technique for telemetry. *Proc. Nat. Telemetering Conf.*, Paper 6.2.

7. Karp, S., and Higuchi, P. K. (1963). An orthogonal multiplexed communication system using modified Hermite polynomials. *Proc. Int. Telemetering Conf., London.*
8. Filipowski, R. F. (1967). Trigonometric product waveforms as the basis of orthogonal sets of signals. *Proc. Nat. Telemetry Conf., Minneapolis, Minnesota,* pp. 36–40.
9. Harmuth, H. F. (1969). Applications of Walsh functions in telecommunications. *IEEE Spectrum* **6** (11), 82–91.
10. Bagdasarjanz, F., and Loretan, R. (1970). Theoretical and experimental studies of a sequency multiplex system. *Proc. Symp. Applic. Walsh Functions, Washington, D.C.,* AD707431, pp. 36–40.
11. Durst, D. I. (1972). Results of multiplexing experiments using Walsh functions. *Proc. Symp. Applic. Walsh Functions, Washington, D.C.,* AD744650, pp. 82–88.
12. Schreiber, H. H. (1973). A review of sequency multiplexing. *Proc. Symp. Applic. Walsh Functions, Washington, D.C.,* AD763000, pp. 18–33.
13. Harmuth, H. F., and Murty, S. S. R. (1973). Sequency multiplexing of digital signals. *Proc. Symp. Applic. Walsh Functions, Washington, D.C.,* AD763000, pp. 202–209.
14. Furrer, F., Shar, A., and Maurer, M. (1972). A Walsh function power cable monitoring system. *Proc. Symp. Applic. Walsh Functions, Washington, D.C.,* AD744650, pp. 89–93.
15. Gordon, J. A., and Barrett, R. (1971). Correlation recovered adaptive majority multiplexing. *Proc. IEE* **118,** 417–422.
16. Gordon, J. A., and Barrett, R. (1972). Group multiplexing by concatenation of non-linear code division systems. *Proc. Symp. Applic. Walsh Functions, Washington, D.C.,* AD744650, pp. 73–81.
17. Mukherjee, A. K., and Mukhopadhyay, D. (1978). A method for increasing the number of majority multiplexed channels. *Proc. IEEE* **66,** 1096–1097.
18. Qishan, Z. (1980). Walsh telemetry systems. *Telemetry Eng.* **1,** (3), 8–11.
19. Changxin, F. (1980). The applications of Walsh functions in the transmission of information. *Ann. Meet. Chinese Electron. Inst. Info. Theory Soc., 3rd, Beijing,* pp. 218–221.
20. Changxin, F. (1980). On even channel majority multiplexing. *IEEE Symp. Electromag. Compat., Baltimore,* pp. 349–353.
21. Zheng, H., and Youwei, Y. (1980). An adaptive multiplex delta modulation system. *IEEE Symp. Electromag. Compat., Baltimore,* pp. 346–348.
22. Jones, H. E. (1972). Bandwidth economy for multiplexed digital signals. *IEEE Int. Conf. Commun. Baltimore,* pp. 36–41.
23. Abramson, N. (1963). "Information Theory and Coding." McGraw-Hill, New York.
24. Ash, R. (1963). "Information Theory," Interscience, New York.
25. Davies, D. W., and Barber, D. L. A. (1973). "Communication Networks for Computers." Wiley, New York.
26. Shannon, C. E. (1948). The mathematical theory of communication. *Bell Syst. Tech. J.* **27,** 379–423.
27. Hamming, R. W. (1950). Error detecting and correcting codes. *Bell Syst. Tech. J.* **26,** 147–160.
28. Elias, P. (1955). Coding for Two Noisy Channels: Third London Symposium on Information Theory. Academic Press, New York.
29. Reed, I. S. (1954). A class of multiple-error correcting codes and the decoding scheme. *IRE Trans. Info. Theory* **IT-4,** 38–49.
30. Peterson, W. W. (1961). "Error Correcting Codes." Wiley, New York.
31. Green, R. R. (1966). A serial orthogonal decoder. Space Programs Summary, Vol. IV, pp. 37–39, 274–225. Jet Propulsion Lab, Pasadena, California.
32. Harmuth, H. F. (1977). "Sequency Theory: Foundations and Applications." Academic Press, New York.

33. Sarry, A. L. (1975). A class of carry-free and mixed operand-operator Walsh-type transforms and some applications. Ph.D. thesis, Catholic University, Washington, D.C.
34. Robinson, G. S. (1972). Quantisation noise considerations in Walsh transform image processing. *Proc. Symp. Applic. Walsh Functions, Washington, D.C.,* AD744650, pp. 240-247.
35. Redinbo, G. R. (1972). Linear mean-square error codes. *Proc. Symp. Applic. Walsh Functions, Washington, D.C.,* AD744650, pp. 330-336.
36. Redinbo, G. R., and Cheung, W. Y. (1982). The design and implementation of unequal error-correcting coding systems. *IEEE Trans. Commun.* **COM-30,** 1125-1135.
37. Pratt, W. K. (1977). "Digital Image Processing." Prentice-Hall, Englewood Cliffs, New Jersey.
38. Harmuth, H. F. (1981). Fundamental limits for radio signals with large relative bandwidth. *IEEE Trans. Electromag. Compat.* **EMC-23,** 37-43.
39. Harmuth, H. F. (1977). Interference caused by additional radio channels using non-sinusoidal carriers. *IEEE Electromag. Compat. Symp., Zurich, Switzerland,* pp. 25-30.
40. Lally, J. F., Hong, Y. K., and Harmuth, H. F. (1974). Experimental transmitter and receiver for electromagnetic Walsh waves. *Proc. Symp. Applic. Walsh Functions, Washington, D.C.,* pp. 176-183.
41. Fralick, S. (1972). Radiation of E.M. waves with Walsh function time variation. *Conf. Rec. Int. Conf. Commun. Philadelphia,* pp. 38-16-38-19.
42. Frank, T. (1971). Circuitry for the reception of Walsh waves. *Proc. Theory Applic. Walsh Functions, Hatfield Polytechnic, England.*
43. Harmuth, H. F. (1974). Electromagnetic Walsh waves: Transmitters, receivers and their applications. *Proc. Symp. Applic. Walsh Functions, Washington, D.C.,* pp. 201-215.
44. Harmuth, H. F. (1978). Radio signals with large relative bandwidth for over-the-horizon radar and spread spectrum communication. *IEEE Trans. Electromag. Compat.* **ENC-20,** 501-512.
45. Harmuth, H. F. (1977). Selective reception of periodic electromagnetic waves with general time variation. *IEEE Trans. Electromag. Compat.* **EMC-19,** 137-144.
46. Davis, J. R., Baker, D. J., Shelton, J. P., and Ament, W. S. (1979). Some physical constraints on the use of 'carrier free' waveforms in radio wave transmission systems. *IEEE Proc.* **67,** 884-891.
47. Lackey, R. B. (1972). The wonderful world of Walsh functions. *Proc. Symp. Applic. Walsh Functions, Washington, D.C.,* AD744650, pp. 2-7.
48. Harmuth, H. F. (1975). Range-Doppler resolution of electromagnetic Walsh waves in radar. *IEEE Trans. Electromag. Compat.* **EMC-17,** 106-111.
49. Griffiths, L. J., and Jacobson, L. A. (1974). The use of Walsh functions in the design of optimum radar waveforms. *Proc. Symp. Applic. Walsh Functions, Washington, D.C.*
50. Woodward, P. M. (1953). "Probability and Information Theory, with Applications to Radar." McGraw-Hill, New York.
51. Rihaczek, A. W. (1971). Radar waveform selection: A simplified approach. *IEEE Trans. Aerospace Electron. Syst.* **AES-7,** 6.
52. Griffiths, L. J. (1973). The extraction of target information from radar signals which use Walsh function modulation formats. *Proc. Symp. Applic. Walsh Functions, Washington, D.C.,* AD763000, pp. 242-247.
53. Cook, J. C. (1960). Proposed monocycle-pulse very high frequency radar for airborne ice and snow measurement. *AIEE Trans. Commun. Electron.* **79,** 588-594.
54. Chapman, J. C. (1976). Experimental results with a Walsh wave radiator. *Nat. Telecom. Conf. Record,* Vol. III, pp. 44.2.1-44.2.3.

55. Crousan, H. M., and Proud, J. M. (1970). Wideband antenna development. Rome Air Develop. Center Tech. NTIS Report AD870224, Washington, D.C.
56. Schmitt, H. J., Harrison, C. W., and Williams, C. S. (1966). Calculated and experimental response of thin cylindrical antennas to pulse excitation. *IEEE Trans. Anten. Propag.* **AP-14,** 120–127.
57. Harmuth, H. F. (1983). Antennas for non-sinusoidal waves I-radiators. *IEEE Trans. Electromag. Compat.* **EMC-25,** 13–24.
58. Zaiping, N. (1983). Radiation characteristics of travelling wave antennas excited by non-sinusoidal current. *IEEE Trans. Electromag. Compat.* **EMC-25,** 24–31.
59. Harmuth, H. F. (1983). On the effect of absorbing materials on electromagnetic waves with large relative bandwidths. *IEEE Trans. Electromag. Compat.* **EMC-25,** 32–39.
60. Spetner, L. M. (1974). Radiation of arbitrary electromagnetic waveforms. *Proc. Symp. Applic. Walsh Functions Sequency Theory,* pp. 249–274.
61. Hussain, M. G. M. (1983). Angular resolution with non-sinusoidal waves. *IEEE Conf. EASCON Washington, D.C.,* pp. 1–11.
62. Hussain, M. G. M. (1983). Signal processing techniques for beam forming with distorted non-sinusoidal waveform. *IEEE Conf. Electromag. Compat., Arlington, Virginia,* pp. 1–6.

Chapter 8

Logical Design and Analysis

8.1 Introduction

Classification and design of digital logic systems have been based in the past on intuitive methods supported by simple application of Boolean algebra and truth tables and limited use of logic package systems. This situation is now changing, and more efficient and comprehensive techniques are beginning to emerge. A major difficulty has been found, for example, in the simplification of Boolean functions. Conventional techniques make use of *diagram methods* such as Veitch and Karnaugh maps and Quine–McCluskey *table methods,* together with other topological and algebraic operations [1–3]. These techniques, and indeed the general methods of Boolean algebra, are not now sufficient to deal with the more powerful logic systems made possible by using CMOS and other LSI technology, particularly where the number of logic variables exceeds six or seven and in multi-input or multi-output logic structures such as PLA and CCD designs. The older techniques are often seen to fail completely in determining all the possible results from complex multi-gate systems and non-canonical structures.

A design philosophy which points a way out of this difficulty is the use of domain transformations for the logical description of digital designs. The application and use of orthogonal functions for this purpose was first suggested by Coleman in 1961 [4]. This arose out of a theoretical study of magnetic core logic in which Fourier transformation of an equivalent Boolean logic formulation allowed the simplification of complex gate structures.

More recent work by Lechner [5] and Karpovsky [6] has established a sound theoretical basis for the many synthesis and design techniques which have derived from this approach. The approach applies principally to the classification and synthesis of logical systems leading to the simplification of the logical circuits to an acceptable minimum.

Intuitively we recognise that bi-valued functions such as those of Walsh and Rademacher are obvious candidates for the transformation process, and considerable development in applying these has been carried out by Hurst [7] and others within the past decade. The techniques of domain analysis has also led to their use in the design of higher logic functions such as *threshold logic gates* [8]. Here a subset of the Walsh coefficients, known as the *Chow parameters* [9], have proved particularly useful. Other combinational logic modules capable of implementing a series of universal logic gates and similar multi-input gate structures may be formulated through the same transformation methods [10]. An extremely interesting by-product of these applications is the possibility of specific design of easily-tested logic — a factor of some importance in the fabrication of complex LSI chips and in the related areas of logical test pattern generation and fault diagnosis for digital systems [11].

In the discussion which will follow some familiarity with Boolean logic analysis at a level found in standard textbooks on this subject is assumed. Since the key to all these newer developments in the use of sequency functions is the transformation process and since this differs from the general methods reviewed earlier, a brief introduction will be given.

8.2 Rademacher–Walsh ordering

As discussed earlier many alternative orderings of the Walsh series can be formulated without affecting their orthogonal properties. Considering only a derivation through an orthogonal matrix, it can be shown [6] that the following two operations can be carried out and orthogonality retained:

1. The signs of all the entries of any number of rows (or columns) may be changed.

2. Any row (or column) may be interchanged with any other row (or column).

However, symmetry may be lost with these operations, so that the inverse transformation will no longer be identical to forward transformation. For this reason very few of these alternative orderings are used. An exception is the Rademacher–Walsh ordering introduced briefly in Subsection 1.3.2. It has been shown that any Walsh series can be derived from the product of several Rademacher series (Eq. (1.37)). This is usually expressed as a Paley-ordered series which is symmetrical in form. A non-symmetrical series can be generated by taking combinations of the Rademacher functions one at a time, two at a time, three at a time, etc., in ascending order. Thus for $N = 2^p = 8$ we can express the series as RAD 0, RAD 1, RAD 2, RAD 3, RAD 1, 2, RAD 1, 3, RAD 2, 3, RAD 1, 2, 3 where the arithmetic products of the Rademacher functions are given as RAD 1, 3 = RAD 1 × RAD 3, etc.

This was shown as a waveform series in Fig. 1.12 but is more conveniently expressed for our purpose as a matrix which for $N = 8$ is given as

$$\mathbf{RW} = \begin{bmatrix} 1 & 1 & 1 & 1 & 1 & 1 & 1 & 1 \\ 1 & 1 & 1 & 1 & -1 & -1 & -1 & -1 \\ 1 & 1 & -1 & -1 & 1 & 1 & -1 & -1 \\ 1 & -1 & 1 & -1 & 1 & -1 & 1 & -1 \\ 1 & 1 & -1 & -1 & -1 & -1 & 1 & 1 \\ 1 & -1 & 1 & -1 & -1 & 1 & -1 & 1 \\ 1 & -1 & -1 & 1 & 1 & -1 & -1 & 1 \\ 1 & -1 & -1 & 1 & -1 & 1 & 1 & -1 \end{bmatrix} \begin{matrix} = \text{RAD } 0 \\ = \text{RAD } 1 \\ = \text{RAD } 2 \\ = \text{RAD } 3 \\ = \text{RAD } 1, 2 \\ = \text{RAD } 1, 3 \\ = \text{RAD } 2, 3 \\ = \text{RAD } 1, 2, 3 \end{matrix} \quad (8.1)$$

If we replace the $+1$ values in this matrix by a logical 0 and -1 by a logical 1, we obtain

$$\mathbf{RW}_b = \begin{bmatrix} 0 & 0 & 0 & 0 & 0 & 0 & 0 & 0 \\ 0 & 0 & 0 & 0 & 1 & 1 & 1 & 1 \\ 0 & 0 & 1 & 1 & 0 & 0 & 1 & 1 \\ 0 & 1 & 0 & 1 & 0 & 1 & 0 & 1 \\ 0 & 0 & 1 & 1 & 1 & 1 & 0 & 0 \\ 0 & 1 & 0 & 1 & 1 & 0 & 1 & 0 \\ 0 & 1 & 1 & 0 & 0 & 1 & 1 & 0 \\ 0 & 1 & 1 & 0 & 1 & 0 & 0 & 1 \end{bmatrix} \begin{matrix} = r_0 \\ = r_1 \\ = r_2 \\ = r_3 \\ = r_{1,2} \\ = r_{1,3} \\ = r_{2,3} \\ = r_{1,2,3} \end{matrix} \quad (8.2)$$

This matrix now has the interesting property that the second to pth row can be used to form all subsequent rows by a process of modulo-2 addition neglecting any carries so formed, viz., $r_{1,2} = r_1 \oplus r_2$, $r_{1,3} = r_1 \oplus r_3$, $r_{2,3} = r_2 \oplus r_3$, $r_{1,2,3} = r_1 \oplus r_2 \oplus r_3$, etc. This is exactly equivalent to the exclusive-OR operation in Boolean logic, and this particular form of coding is relevant to the Boolean synthesis of digital networks. Thus we find that the first $p + 1$ variables may constitute the actual binary input to a digital system, whilst the remaining exclusive-OR variables can represent all possible modulo-2 additions of the first $p + 1$ variables, here called the primary set of variables. The matrix expression of Eq. (8.2) is known as the *binary Walsh matrix* $\mathbf{RW}_b$ and was first suggested by Lechner [5] as a method of expressing binary logic functions.

Whilst there are considerable advantages in synthesising logical systems in terms of the $\mathbf{RW_b}$ matrix alone, it should be noted that this constrains the synthesis to the manipulation of exclusive-OR relationships and the independent input variables. Through suitable application of the row and column interchange operations described earlier, it is possible to arrive at an orthogonal matrix which does not allow spectral design techniques of the Rademacher–Walsh type, which favour, say, NAND realisations rather than any other logic element building block [12]. This procedure may be less efficient than the Rademacher–Walsh transformation but would be of value in consideration of vertex-based logic, i.e., logic based entirely on AND/OR or NAND/NOR realisations.

8.2.1 Transformation

The transformation of a set of input variables x_i can be obtained by matrix multiplication by $\mathbf{RW}$ or $\mathbf{RW_b}$ to yield a set of N spectral coefficients, the Rademacher–Walsh spectrum R_i. To achieve this with the matrix $\mathbf{RW}$ we can simply use a fast Walsh transform algorithm and then reorder the results to give Rademacher–Walsh order [13]. In this case, however, $\mathbf{RW} \neq \mathbf{RW}^{-1}$ and a separate inverse transformation algorithm is needed. This inverse matrix $\mathbf{RW}^{-1}$ is derived simply by interchanging rows with columns of $\mathbf{RW}$, viz.,

$$\mathbf{RW}^{-1} = \begin{bmatrix} 1 & 1 & 1 & 1 & 1 & 1 & 1 & 1 \\ 1 & 1 & 1 & -1 & 1 & -1 & -1 & -1 \\ 1 & 1 & -1 & 1 & -1 & 1 & -1 & -1 \\ 1 & 1 & -1 & -1 & -1 & -1 & 1 & 1 \\ 1 & -1 & 1 & 1 & -1 & -1 & 1 & -1 \\ 1 & -1 & 1 & -1 & -1 & 1 & -1 & 1 \\ 1 & -1 & -1 & 1 & 1 & -1 & -1 & 1 \\ 1 & -1 & -1 & -1 & 1 & 1 & 1 & -1 \end{bmatrix} \quad (8.3)$$

We will, of course, need to divide the output by $1/2^p$ to achieve correct scaling.

The recoding of the matrix $\mathbf{RW}$ to form the $\mathbf{RW_b}$ matrix is obvious, but there is another way of achieving a transformation of $x_i(b) \cdot \mathbf{RW_b}$, where $x_i(b)$ is in binary form. It can be shown [8] that the spectral coefficients

$R_i(i = 0, 1, \ldots, N-1) = \{$number of agreements between the values of 0 or 1 in the row r_i of the matrix and the input vector $x_i(b)\}$

$\quad - \{$number of disagreements$\}$ \hfill (8.4)

As an example, if we let $x_i(b) = 0, 1, 1, 0, 1, 1, 1, 0$ and compare this with the first row of Eq. 8.2, $r_0 = 0, 0, 0, 0, 0, 0, 0, 0$, we have (3 agreements − 5 disagreements) $= -2 = R_0$. Similarly for r_1 we have (5 agreements $-$ 3 disagreements) $= 2 = R_1$. Carrying this out for all the spectral coefficients of x_i results in $R_i = -2, 2, -2, -2, 2, 2, 6, 2$, forming the complete Rademacher–Walsh spectrum. Note that the magnitude of R_i indicates a measure of agreement between $x_i(b)$ and r_i and therefore expresses a correlation between the input function $x_i(b)$ and the rows of the $\mathbf{RW_b}$ matrix. Hence the spectral coefficient values represent the correlation coefficients and indicate how 'like' $x_i(b)$ to each row of the transform matrix is. If $x_i(b)$ is exactly the same as one of the rows of $\mathbf{RW_b}$, then the resulting spectral coefficient value will be $\pm 2^p$ (± 8 in our example) and *all* other coefficients zero.

A hardware mechanism to carry this out as part of a logic fault detection system is described by D'Alton [14].

8.3 Synthesis of digital networks

The purpose of deriving the Rademacher–Walsh transform of a Boolean function, instead of using the more familiar truth table and mapping notations, is to simplify certain operations which are difficult or not even possible where the representation is limited to only two binary values. A Rademacher–Walsh transform of a Boolean function can be derived by first expressing the function as a truth table. Consider the Boolean function $f(x) = x_1 \bar{x}_2 + x_2 \bar{x}_3$ corresponding to the logic circuit and truth table given in Fig. 8.1. The resulting vector $F(x)$ is obtained as $F(x) = 0, 0, 1, 0, 1, 1, 1, 0$, which when transformed into the sequency domain through the Rademacher–Walsh transform gives $R_i = 0, 4, 0, -4, 4, 0, 4, 0$. The important factor to note here is that the 2^p values of R_i completely define or classify the logic function in a unique way, as does a Boolean truth table but in a dissimilar manner. Also as a direct result of Rademacher–Walsh ordering the first $p + 1$ spectral coefficients relate to the logical inputs x_0 to x_p, whilst the remainder relate to modulo-2 additions of the inputs. These first $p + 1$ spectral coefficients are the Chow parameters, and for certain threshold logic functions it has been shown [8] that only these parameters are actually needed to classify uniquely a Boolean function (see also Muzio *et al.* [15]). Where the Chow parameters contain the highest-magnitude components, the function proves easiest to synthesise. This can be understood if we consider a function whose spectrum consists of only one positive component at, say, R_2. This function will then correlate exactly with the x_2 input and the function synthesised simply by connecting the input x_2 with the output. If, however, this component is situated at $R_{1,2}$ we see from Eq. (8.2) that the synthesis would be implemented by an exclusive-OR gate connected between x_1, x_2 and the output. Similarly functions which exhibit strong spectral characteristics in the second-order region, where the modulo-2 addition of two first-order terms is relevant, are easier to synthesise than in the third-

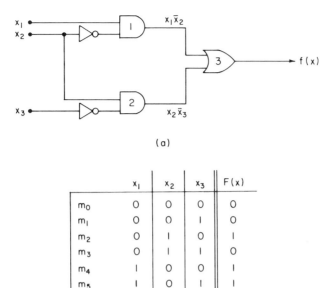

FIG. 8.1. (a) Boolean logic circuit for $f(x) = (x_1\bar{x}_2 + x_2\bar{x}_3)$; (b) truth table.

order region, where modulo-2 addition of three first-order terms is involved, and so on.

For this reason considerable interest has been taken in those digital synthesis methods which involve higher-order logic, such as threshold gates (which are specified uniquely by only the first $p + 1$ spectral coefficients) and exclusive-OR gates (which carry out the modulo-2 addition of two functions).

Whilst the application of Rademacher–Walsh spectral methods may be applied to the simplification of vertex logic structures with some advantage [16], its major impact has been in the use of more powerful and complex gate structures and non-canonical forms in which not all the logic inputs are able to be specified. Before we consider how spectral analysis is applied to the simplification of Boolean logic systems, we first need to consider the role of these complex logic gates.

8.3.1 Threshold logic

The vertex gates consisting of a single AND/OR or NAND/NOR operation are responsive to only a single minterm of a logic function and may, therefore, be considered the least efficient of the possible binary logic structures. The exclusive-OR/NOR permits a more complex operation to be carried out on the basis of modulo-2 addition of a number of inputs and hence minterms. A greater degree of compactness in carrying out logic functions is possible with multi-input gates with which an element of logic input selection can be carried out. One example is the majority logic gate with which the logical output assumes the logical state found in the majority of its inputs,

$$f(x) = 1 \quad \text{if} \quad \sum_{i=1}^{n} x_i \geq n/2 \quad \text{and} \quad 0 \text{ otherwise} \quad (8.5)$$

where the number of inputs i is generally taken to be an odd number.

Threshold gates are similar, but each gate input can be *weighted* arithmetically and summed to give an output if certain preset thresholds are exceeded [17]. Thus, referring to Fig. 8.2, if $A, B, \ldots, N$ are input values (0 or 1), $a, b, \ldots, n$ are the multiplying weights (real positive integers) and t_1, t_2 are the upper and lower threshold levels (real positive integers), we can express the gate operation as

$$\begin{aligned} z &= 1 \quad \text{if} \quad (aA + bB + \cdots nN) \geq t_1 \\ z &= 0 \quad \text{if} \quad (aA + bB + \cdots nN) \leq t_2 \end{aligned} \quad (8.6)$$

Since the first $p + 1$ Rademacher–Walsh spectral coefficients are sufficient to characterise uniquely a threshold function the use of such functions can simplify logic synthesis. Tables of spectral coefficients together with the weights necessary to define the functions are available up to $p = 7$ [18], and the use of such tables is illustrated by Edwards [19]. Two useful operations on Boolean functions, namely, *spectral translation* and *disjoint spectral translation*, have also been described. These allow the zero-ordered spectral coefficient of any Boolean function to be interchanged with any first-ordered spectral coefficient [19, 20]. By repeating this operation it is possible to synthesise any Boolean function, either directly or through an equivalent function of a similar class whose synthesis is known (see Subsection 8.5.2). The power of this form of Boolean classification is apparent from the tables generated by Edwards [19]. For $p \leq 4$ the total number of possible Boolean functions is given as 65,536. This may be reduced to 8, thus showing that

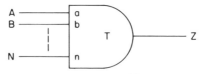

FIG. 8.2. A threshold gate.

only 8 unique logic modules of which 7 are threshold functions, are necessary to synthesise this range of Boolean functions. The small number of higher-logic gates necessary to synthesise a wide range of Boolean logic has prompted designs of universal threshold gates capable of implementing any combinational Boolean function up to $p = 5(10)$ multi-output gates [21] and multi-threshold logic gates [22]. This latter is relevant to CCD technology in which large-scale array processing may be carried out.

8.4 Minimisation of logic functions

Early methods of minimisation of complex logic functions to reduce the number of logic elements required in a practical design consisted of a gradual reduction towards the prime implicants. This was achieved through algebraic or topological methods commencing with the specification of the 2^p discrete values (minterms) of the Boolean function. As discussed earlier the values of the minterms could also be mapped uniquely into a sequency spectrum and synthesis methods evoked to produce a smaller set of higher-logic functions representing the required logic relationships. The number of input variables for many logic requirements is, however, increasing. For instance, PLA units typically have 14 inputs. Further, many problems involve non-canonical logic in which there are a relatively large number of undefined ('don't care') positions. Optimum design for minimisation in these cases means computer-aided design directed to the generation of prime implicants. Since the problems often exhibit large numbers of variables and/or prime implicants, approximate solutions have to be sought.

Apart from the well-known algebraic or topological methods (e.g., Karnaugh maps or Quine–McCluskey techniques), there are several transformation techniques which can be applied. One method is to examine the spectrum for detectable symmetries between the coefficients which relate to the input variables. This can lead to the identification of those minterms which may be omitted from the logic synthesis procedure and can be extended to the synthesis of non-canonical (incomplete logic) functions.

8.4.1 Detection of symmetries

Some form of symmetry is apparent in the input–output relationship of every Boolean function, which, when recognised, may lead to a more efficient function synthesis. Hurst [23] defines two forms of symmetry:

1. In *a totally symmetric function* $f(x)$ any permutation of its input variables leaves the function output invariant. Thus an AND function $f(x) = x_1 x_2 x_3$ can be rearranged as $f(x) = x_3 x_2 x_1$, etc., and remain invariant.

8 Logical Design and Analysis

2. *In a partially symmetric function* $f(x)$ interchanges within one or more sets of m input variables $2 \leq m < p$ within a total of p variables leave the function invariant. Thus $f(x) = x_1 x_2 + x_3$ is partially symmetric in x_1 and x_2 since interchanging these has no effect on the output.

Cases of total symmetry are not of great interest in minimisation of Boolean logic since this form of symmetry is frequently apparent from the definition of a function, e.g., majority gates, NAND gates, a full adder function. What is more difficult to detect are the various forms of partial symmetry that may be present between two input variables x_i and x_j of a complex function $f(x)$.

Partial symmetry can be defined as

$$f(x_1, \ldots, x_i, x_j, \ldots, x_p) \equiv f(x_1, \ldots, x_j, x_i, \ldots, x_p) \quad (8.7)$$

where $i \neq j$ and the expression includes the complemented forms $\bar{x}_i, \bar{x}_j$. Thus under certain conditions we can replace $x_i x_j$ by $x_j x_i$ or by its complemented form and possibly achieve a simplification in the hardware formulation of the function. Also, since we are basing this on a spectral evaluation, it is likely that the resulting simplification will be easier to test since it may include a higher complement of exclusive-OR gates. A common method of identifying symmetry is through recognition of identical columns of a Karnaugh map and replacing one of these columns by 'don't care' logic values, thus permitting more design latitude in the allocation of values for the input minterms. An earlier method employed algebraic expansion to identify pairs of symmetric functions. These methods proved cumbersome and difficult to apply for large numbers of minterms.

The spectral approach developed by Hurst and others [11, 23] consists of a series of simple checks and comparisons with pairs of spectral coefficients. Only an outline of the method will be given here. For a full description and proofs the reader is referred to the works already cited.

Two basic types of partial symmetry can be identified, *equivalence symmetry* (ES), where

$$x_i x_j = 0, 0 \quad \text{and} \quad 1, 1 \quad (8.8a)$$

and *non-equivalent symmetry* (NES), where

$$x_i x_j = 0, 1 \quad \text{and} \quad 1, 0 \quad (8.8b)$$

Several other forms of symmetry are defined by Hurst [7], but in the interest of simplicity will not be discussed here.

Examples of these two forms of symmetry are given in Fig. 8.3a,b. In the first case the columns for $x_1 x_2 = 0, 0$ and $1, 1$ are identical, indicating the presence of equivalence symmetry, written ES(x_1, x_2). In the second case the

8.4 Minimisation of Logic Functions

x_3x_4 \ x_1x_2	0,0	0,1	1,1	1,0
0,0	1	0	1	1
0,1	0	1	0	1
1,1	0	0	0	0
1,0	0	1	0	0

(a)

x_3x_4 \ x_1x_2	0,0	0,1	1,1	1,0
0,0	0	1	0	1
0,1	0	0	1	0
1,1	0	1	0	1
1,0	1	0	1	0

(b)

FIG. 8.3. (a) Equivalence symmetry in x_1, x_2; (b) non-equivalence symmetry in x_1, x_2.

columns for $x_1 x_2 = 0, 1$ and $1, 0$ are identical, indicating the presence of non-equivalence symmetry, written NES(x_1, x_2).

To detect these and other forms of symmetry a Rademacher–Walsh transformation of the vertices of the logical function is taken and various selected pairs of coefficients compared.

Consider first a search for equivalence symmetry in, say, x_1 and x_2. A first check (necessary but not sufficient by itself) for ES(x_1, x_2) is

$$R_1 + R_2 = 0 \qquad (8.9)$$

indicating symmetry between input variables x_1 and x_2. This is termed the primary check. Sufficiency is not yet proved since the 0 and 1 valued minterms in areas $\bar{x}_1 \bar{x}_2$ and $x_1 x_2$ can be in any order. Further secondary tests are necessary with the remaining pairs of spectral coefficients containing x_1 or x_2, viz.,

$$R_{1,2} + R_{2,3} = 0$$
$$R_{1,4} + R_{2,4} = 0 \qquad (8.10)$$
$$R_{1,3,4} + R_{2,3,4} = 0$$

Similarly the tests for non-equivalence symmetry in x_1, x_2, namely, NES(x_1, x_2) are

$$R_1 - R_2 = 0 \qquad (8.11)$$

and

$$R_{1,3} - R_{2,3} = 0$$
$$R_{1,4} - R_{2,4} = 0 \qquad (8.12)$$
$$R_{1,3,4} - R_{2,3,4} = 0$$

Essentially similar checks may be carried out in a search for other types of symmetries. As an example a four-variable function

$$f(x) = \bar{x}_1\bar{x}_2(x_3 + x_4) + x_3 x_4(\bar{x}_1 + \bar{x}_2)$$
$$+ \bar{x}_3\bar{x}_4(\bar{x}_1 x_2 + x_1 \bar{x}_2) + x_1 x_2 x_3 \bar{x}_4 \tag{8.13}$$

is mapped in Fig. 8.4. The spectrum of the vertices may be calculated as

$$\begin{array}{cccccccc} R_0 & R_1 & R_2 & R_3 & R_4 & R_{1,2} & R_{1,3} & R_{1,4} & R_{2,3} \\ -2 & -4 & -4 & 4 & 0 & 0 & 0 & 4 & 0 \end{array}$$

$$\begin{array}{ccccccc} R_{2,4} & R_{3,4} & R_{1,2,3} & R_{1,2,4} & R_{1,3,4} & R_{2,3,4} & R_{1,2,3,4} \\ 4 & -4 & 4 & 0 & 0 & 0 & 12 \end{array} \tag{8.14}$$

The primary tests (Eqs. (8.9) and (8.11)) reveal

$$\text{NES}(x_1, x_2) \quad \text{since} \quad R_1 - R_2 = 0$$
$$\text{ES}(x_1, x_3) \quad \text{since} \quad R_1 + R_3 = 0$$
$$\text{ES}(x_2, x_3) \quad \text{since} \quad R_1 + R_3 = 0$$

The secondary tests (Eqs. (8.10) and (8.12)),

$$R_{1,3} - R_{2,3} = 0, \quad R_{1,4} - R_{2,4} = 0 \quad \text{and} \quad R_{1,3,4} - R_{2,3,4} = 0$$
$$R_{1,2} + R_{2,3} = 0, \quad R_{1,4} + R_{3,4} = 0 \quad \text{and} \quad R_{1,2,4} + R_{2,3,4} = 0$$
$$R_{1,2} + R_{1,3} = 0, \quad R_{2,4} + R_{3,4} = 0 \quad \text{and} \quad R_{1,2,4} + R_{1,3,4} = 0$$

confirm that these symmetries are valid and may be used in a simplified synthesis of the function.

Note that whilst the NES(x_1, x_2) is easily seen from Fig. 8.4, the two equivalent symmetries in x_1, x_3 and x_2, x_3 are not immediately obvious from the function mapping.

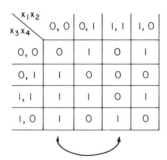

FIG. 8.4. A symmetry example.

8.4.2 Spectral addition

A computer-aided design synthesis due to Lloyd [24] carries out a prime implicant extraction, not by symmetries but by searching the Rademacher–Walsh spectrum looking for selected subsets of coefficients which together add to a maximum value of $2^{\pm p}$. This can be shown to result in the identification of prime implicants related to the position of these coefficients in the spectrum.

Consider first a subset consisting of pairs of adjacent coefficient values. Referring to the Rademacher–Walsh matrix **RW** given in Eq. (8.1), if we add any pair of adjacent rows RAD(a) and RAD(b), then we note that half of the resultant elements are always zero and the remainder take on a value of ± 2. When this summation vector is multiplied by the set of equivalent truth table vertices F, in which the 0 and 1 terms in $F(x)$ are replaced by $+1$ and -1, we obtain the Rademacher–Walsh spectrum $R = R_a + R_b$. But only half the spectral coefficients will have a finite value, namely, ± 2. We may also note that the maximum value of R must be $\pm 2^p$ and will only occur when the ± 1 values in the truth table vertices F correlate exactly with the same valued entries in a and b.

However, since

$$(\text{RAD}(a) \pm \text{RAD}(b))F = (\text{RAD}(a))F \pm (\text{RAD}(b))F$$
$$= R_a \pm R_b = R \qquad (8.15)$$

then R is defined simply as the addition or subtraction of adjacent pairs of spectral coefficients. Since only half of the constituent coefficients are present, $R_a + R_b$ is shown to be the summation of half the possible minterms contained in F.

Now if we consider, initially, only those $R_a + R_b$ summations which result in $\pm 2^p$, the maximum possible values, then this will serve to identify half the possible minterms in the Boolean function.

We consider as an example a three-variable ($p = 3$) function

$$f(x) = \bar{x}_1 x_2 + x_1 \bar{x}_2 + x_2 x_3$$

from which a truth table can be derived, which, when recoded in ± 1 form is

$$F = 1, 1, -1, -1, -1, -1, 1, -1$$

The products R_1 and $R_{2,3}$ will be from Eq. (8.1)

$$R_1 = (\text{RAD } 1)F = m_0 + m_1 + m_2 + m_3 - m_4 - m_5 - m_6 - m_7 \qquad (8.16)$$

and

$$R_{2,3} = (\text{RAD } 2, 3) = m_0 - m_1 - m_2 + m_3 + m_4 \quad m_5 \quad m_6 + m_7 \qquad (8.17)$$

where $m_0, \ldots, m_7$ are the true and false minterms contained in $f(x)$ with $m_0 = \bar{x}_1 \bar{x}_2 \bar{x}_3$, $m_1 = \bar{x}_1 \bar{x}_2 x_3$, etc. (where a zero-entry value for x is taken as $\bar{x}$ (see Fig. 8.1)).

Hence

$$R_1 + R_{2,3} = 2m_0 + 2m_3 - 2m_5 - 2m_6 \qquad (8.18)$$

If now we find that $R_1 + R_{2,3} = +8$, then the value of the individual minterms must be as shown in the Karnaugh representation given in Fig. 8.5a, and this single summation has fully defined half the minterms of $f(x)$. Similarly if, say, $R_1 - R_2 = -8$, then we would have $2_{m_2} + 2_{m_3} - 2_{m_4} - 2_{m_5} = -8$, again defining fully half the possible minterms (Fig. 8.5b).

Further subsets of spectral coefficients can be taken four at a time, eight at a time, etc., to yield other sets of minterm values to enable the complete synthesis of a given logic function to be realised [25]. Lloyd also shows that the *true* and *false minterms* explicitly defined by a given summation are subject to the following rules:

1. The true (logic 1) minterms of $f(x)$ are those minterms given by the AND of the functions defined by the appropriate rows in the transform matrix **RW**, e.g., x_1 for RAD 1, $x_2 \oplus x_3$ for RAD 2, 3, etc.

2. The false (logic 0) minterms of $f(x)$ are those minterms given by the OR of the same functions defined in **RW**.

Taking the second example of Fig. 8.5b, $R_1 - R_2 = -2^p$, where R_1 correlates with the transform row function x_1 and R_2 correlates with the transform function x_2. Since a summation to -2^p is equivalent to taking the complement of the row functions, we can write $-R_1 + R_2 = 2^p$, which corresponds with the AND of $\bar{x}_1$ and x_2 from rule 1. Also from rule 2 we see that the false

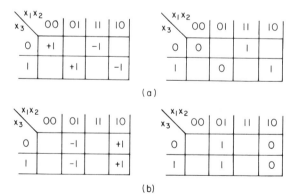

FIG. 8.5. (a) Karnaugh map designations for $R_1 + R_{2,3}$; (b) Karnaugh map designations for $R_1 - R_2$.

8.4 Minimisation of Logic Functions

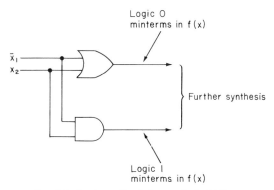

FIG. 8.6. Logic synthesis by spectral addition.

minterms correspond with the OR of $\bar{x}_1$ and x_2. The pertinent section of the derived logic is shown in Fig. 8.6.

Further synthesis to determine the remaining minterms proceeds in a series of stages in which a search for large spectral coefficient summations is made and the appropriate logic interconnections determined. It may be noted that all the true and false minterms that have been synthesised at an earlier stage can, if desired, be made 'don't care' minterms for subsequent stages of the realisation. The problem of the optimum allocation of values 0 or 1 in these minterms so as to give a 'good' or 'best' synthesis has not yet been completely solved in the general case. A detailed flow diagram and description of the complete procedure is given Lloyd and Hurst [26], who point out that, contrary to earlier non-spectral prime implicant extraction algorithms, this procedure requires fewer spectral coefficients to be considered the larger the number of implicants realised at a particular stage in the synthesis. Hence the method is well suited to the synthesis of the larger size of design problems met with in LSI fabrication.

8.4.3 Extraction of all prime implicants

An alternative technique has been developed by Besslich [27, 28] from the prime implicant detection theorem of Lechner [5]. This technique is also based on the summation of sets of spectral coefficients to achieve a maximum value of $\pm 2^p$. In this case, however, an exhaustive algorithmic search in the spectral domain is carried out for all possible 3^p implicants of a given function $f(x)$. Here a matrix summation carries out 2^p summations at a time, and the procedure repeated with further groups of 2^p summations until all prime implicants are extracted.

Commencing with a list of minterms specifying a given Boolean function, the minimisation goal is to synthesise a new and simpler set of minterms consisting only of those irreducible terms that fail to match among themselves in any of their constituent values (literals). In effect, what the technique is doing is to list all possible minimising combinations of the type $Ax + A\bar{x} = A$ which are maximal in the sense that each includes as many of the original minterms as possible. Having done this, we need to identify those prime implicants which are essential to the required function, i.e., contain a logic term not included or *covered* in any other implicant from those which are preferred out of the remaining implicants by reason of their logical content. All other than essential and preferred implicants may then be excluded from the final minimal Boolean expression.

The processes of prime implicant extraction and selection for minimal cover are two separate operations necessary to completely reduce a Boolean expression. Only the process of all prime implicant extraction by sequency transform will be considered here. Methods for the determination of minimal cover, also using transform methods, are described by Besslich [29, 30].

A difficulty with the extraction of all the prime implicants, rather than a selected subset of them as discussed in the preceding section, is the large number of implicants that can occur for numbers of variables greater than about 6. Even a completely specified 6-variable Boolean function may have up to 90 selected prime implicants [31] and, accordingly, hundreds of irredundant forms. Since many multi-input and multi-output logic systems may have an incompletely specified logic form including a number of 'don't care' inputs, the problem becomes a complex one. Hence computer-aided design for prime-implement extraction is essential, and it is here that multi-variable problems benefit most from transform methods that can be implemented by computer program.

The computer-aided technique developed by Besslich is outlined below. Readers are directed to Besslich's 1978 [29] and 1979 [28] works for full details and mathematical proofs for this method.

As in other spectral methods the initial input information consists of the truth table realisation for the Boolean function. However, since the method is equally applicable to non-canonical forms, an incompletely defined Boolean function may be considered. In this example, for which I am indebted to Professor Besslich, the function consists of only 6 of the possible combinations of $p = 4$ variables x_0, x_1, x_2, x_3 shown in boldface type in Fig. 8.7. These include two true minterms, 0000 and 0111 and four false minterms 0001, 1100, 1101 and 1110. The remaining values may be tentatively regarded as resulting in a logical 1 to give

$$F = 1, -1, 1, 1, 1, 1, 1, 1, 1, 1, 1, 1, -1, -1, -1, 1 \qquad (8.19)$$

8.4 Minimisation of Logic Functions

x_0	x_1	x_2	x_3	$f(x)$	$f'(x)$
0	0	0	0	1	1
0	0	0	1	0	-1
0	0	1	0	1	1
0	0	1	1	1	1
0	1	0	0	1	1
0	1	0	1	1	1
0	1	1	0	1	1
0	1	1	1	1	1
1	0	0	0	1	1
1	0	0	1	1	1
1	0	1	0	1	1
1	0	1	1	1	1
1	1	0	0	0	-1
1	1	0	1	0	-1
1	1	1	0	0	-1
1	1	1	1	1	1

FIG. 8.7. Truth table for a non-canonical function.

This is transformed by the naturally-ordered Hadamard transform, viz.,

$$X_i = \mathbf{F} \cdot \mathbf{H}_{16} = 8, 0, -4, 4, 4, 4, 0, 0, 4, 4, 0, 0, -8, 0, -4, 4 \quad (8.20)$$

also coded logical $0 = -1$ and logical $1 = +1$. Note that this is not the same ordering as used in the earlier part of this chapter since the Hadamard, and not the Rademacher–Walsh, transformation matrix is used.

To obtain all possible implicants of $f(x)$ each true (logic 1) minterm of the given function is selected in turn and all product terms which involve this minterm are then determined. This is carried out through a matrix operation on X_i by using a particular kind of Rademacher–Walsh transform shown in Fig. 8.8, termed here the Besslich Rademacher–Walsh matrix $\mathbf{BW}$. In order to generate all the prime implicants involving a given minterm it is necessary first to multiply Eq. (8.20) with the signs of the elements found in the appropriate row in the Hadamard matrix (e.g., row 0 for m_0, row 1 for m_1, etc). For the first minterm 0000 then, this consists of 16 logic ones (from Eq. (1.45)) remembering that $\mathbf{H}_{16} = \mathbf{H}_8 \times \mathbf{H}_2$, so that the signs of X_i remain unchanged, i.e., $X'_i = X_i$. Carrying out the transformation X'_i we have

$$P_{0000}(x) = X'_{0000}\mathbf{BW}_{16} = 16, 16, 16, 16,$$
$$0, 8, 16, 8, 16, 8, 8, 8, 4, 12, 12, 8 \quad (8.21)$$

The implicants are identified according to the maximum value 2^p, which in this case for $p = 4$ is 16. This will, however, include the non-prime impli-

$$BW = \begin{bmatrix} 1 & 1 & 1 & 1 & 1 & 1 & 1 & 1 & 1 & 1 & 1 & 1 & 1 & 1 & 1 & 1 \\ 1 & 1 & 1 & 1 & 1 & 1 & 1 & 1 & 0 & 0 & 0 & 0 & 0 & 0 & 0 & 0 \\ 1 & 1 & 1 & 1 & 0 & 0 & 0 & 0 & 1 & 1 & 1 & 1 & 0 & 0 & 0 & 0 \\ 1 & 1 & 0 & 0 & 1 & 1 & 0 & 0 & 1 & 1 & 0 & 0 & 1 & 1 & 0 & 0 \\ 1 & 0 & 1 & 0 & 1 & 0 & 1 & 0 & 1 & 0 & 1 & 0 & 1 & 0 & 1 & 0 \\ 1 & 1 & 1 & 1 & 0 & 0 & 0 & 0 & 0 & 0 & 0 & 0 & 0 & 0 & 0 & 0 \\ 1 & 1 & 0 & 0 & 1 & 1 & 0 & 0 & 0 & 0 & 0 & 0 & 0 & 0 & 0 & 0 \\ 1 & 0 & 1 & 0 & 1 & 0 & 1 & 0 & 0 & 0 & 0 & 0 & 0 & 0 & 0 & 0 \\ 1 & 1 & 0 & 0 & 0 & 0 & 0 & 0 & 1 & 1 & 0 & 0 & 0 & 0 & 0 & 0 \\ 1 & 0 & 1 & 0 & 0 & 0 & 0 & 0 & 1 & 0 & 1 & 0 & 0 & 0 & 0 & 0 \\ 1 & 0 & 0 & 0 & 1 & 0 & 0 & 0 & 1 & 0 & 0 & 0 & 1 & 0 & 0 & 0 \\ 1 & 1 & 0 & 0 & 0 & 0 & 0 & 0 & 0 & 0 & 0 & 0 & 0 & 0 & 0 & 0 \\ 1 & 0 & 1 & 0 & 0 & 0 & 0 & 0 & 0 & 0 & 0 & 0 & 0 & 0 & 0 & 0 \\ 1 & 0 & 0 & 0 & 1 & 0 & 0 & 0 & 0 & 0 & 0 & 0 & 0 & 0 & 0 & 0 \\ 1 & 0 & 0 & 0 & 0 & 0 & 0 & 1 & 0 & 0 & 0 & 0 & 0 & 0 & 0 & 0 \\ 1 & 0 & 0 & 0 & 0 & 0 & 0 & 0 & 0 & 0 & 0 & 0 & 0 & 0 & 0 & 0 \end{bmatrix} \begin{matrix} = r'_0 \\ = r'_1 \\ = r'_2 \\ = r'_3 \\ = r'_4 \\ = r'_{1,2} \\ = r'_{1,3} \\ = r'_{1,4} \\ = r'_{2,3} \\ = r'_{2,4} \\ = r'_{3,4} \\ = r'_{1,2,3} \\ = r'_{1,2,4} \\ = r'_{1,3,4} \\ = r'_{2,3,4} \\ = r'_{1,2,3,4} \end{matrix}$$

FIG. 8.8. The Besslich Rademacher–Walsh matrix for $N = 16$.

cants which can be shown to occupy the first four positions. Excluding these, the remaining prime implicants are $P_{6.1} = (-0 -0)$ and $P_{8.1} = (0- -0)$. These are the prime implicants containing the first minterm. The implicants of the term 0111 are obtained similarly, but in this case X_i must first be multiplied by the signs of binary row 0111 of the Hadamard matrix to obtain

$$X'_{0111} = 8, 0, 4, 4, -4, 4, 0, 0, 4, 4, 0, 0, 8, 0, -4, -4 \quad (8.22)$$

Hence

$$P_{0111}(x) = X'_{0111} \mathbf{BW}_{16} = 16, 16, 16, 16,$$
$$16, 16, 8, 8, 8, 16, 16, 8, 12, 4, 12, 8 \quad (8.23)$$

Deleting the non-prime implicants (the first five) indicates the remaining prime implicants as

$$P_{5 \cdot 2} = (- -11), \quad P_{9 \cdot 2} = (0-1-) \quad \text{and} \quad P_{10 \cdot 2} = (01- -)$$

There are as many of these transformation-inverse transformations required to complete the process as there are true minterms in F. The true minterms cannot, of course, exceed 2^p for a p-variable problem. In a subsequent paper [32] a fast transform implementation of this method is given in which the forward Hadamard transformation and multiplication by the matrix **BW** are combined as one operation. The problem of final determination of an irredundant prime implicant cover for $f(x)$ using these results is described by Besslich [29].

8.5 Fault diagnosis

We have seen how a given logical function is defined quite uniquely by its spectral coefficients. These collectively form a kind of spectral *signature* of a logic function and the same signature cannot be obtained from any other function. It is not hard to understand, therefore, that any malfunction of the hardware related to this function will cause it to behave differently and a change will be seen in its spectral characteristics. Further it is apparent that the relative magnitude of these coefficients compared with those derived from a hardware unit functioning properly should indicate the importance of the particular input parameter in determining the spectral values. This is shown particularly by the Rademacher–Walsh transformation through its correlation properties (Subsection 8.2.1).

Consider, for example, the logic circuit given in Fig. 8.1 for the Boolean function $f(x) = x_1 \bar{x}_2 + x_2 \bar{x}_3$. The spectral characteristic for this is shown in Fig. 8.9a. The relative magnitude of each coefficient is a measure of the importance of a particular input parameter x_i ($i \neq 0$) in determining the logical value of the network output. A high positive value indicates a high degree of dependence of the output on a particular x_i input and a high negative value on the complement of the value $\bar{x}_i$. A maximum-valued coefficient (i.e., $\pm 2^p$) indicates complete dependence, and other inputs are ineffective.

In this example we can state that the output of the logic network is highly dependent on RAD 1 and hence input x_1, highly dependent on RAD 3 and hence input $\bar{x}_3$ (since the dependence is a negative one) and further highly dependent on the Boolean combinations of $x_1 \oplus x_2$ and $x_2 \oplus x_3$ related to RAD 1, 2 and RAD 2, 3, which occur within the network. This last dependence is important since it can indicate how a fault occurring within a logic network may be determined with only input/output access permissible (e.g., a complex LSI chip).

An example is given by Bennetts and Hurst [11], who demonstrate this possibility applied to the simple logic network we have been discussing (Fig. 8.1). Consider that the upper input of gate 2 becomes *stuck-at-1* (s.a.1) so that the network behaves as if it carries out the logic function $f'(x) = x_1 \bar{x}_2 + \bar{x}_3$. The output spectrum, shown in Figure 8.9b, indicates that a prominent peak for RAD 3 relating to $\bar{x}_3$ is now present. This is sufficient by itself to identify the s.a.1 node in the network as one associated with gate 2 in Fig. 8.1a.

It is clear that, since each individual logic network has its own unique spectral signature, an effective change in the network caused by a specific fault occurring anywhere within the network will also have its characteristic spectral signature. Hence it is possible to compile a fault 'dictionary' giving a tabular list of possible faults associated with a given spectrum and thus aid fault diagnosis.

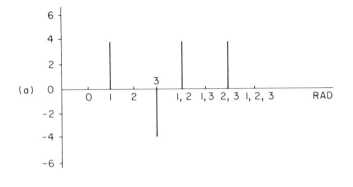

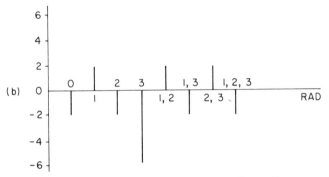

FIG. 8.9. (a) Rademacher–Walsh spectrum for $f(x) = (x_1\bar{x}_2 + x_2\bar{x}_3)$; (b) Rademacher–Walsh spectrum for $f(x) = (x_1\bar{x}_2 + \bar{x}_3)$.

In many practical circumstances it may be time consuming to have to consider or compare all the spectral coefficients, and attempts have been made to identify a fault condition with fewer than the total number of coefficients identified for a given logic function. The Chow parameters have been suggested for this purpose by Hurst [33] and indeed have been proved all that is necessary in those logic functions which can be realised from a threshold core logic of the type discussed earlier in Subsection 8.3.1 [15]. This subset of the Rademacher–Walsh coefficients is, however, not sufficient for functions lying outside this threshold class, and other more general methods of logic testing may prove more useful.

8.5.1 Test pattern generation

A general method which can be used to determine a single s.a.1 or s.a.0 *(struck-at-0)* fault in a combinational network is to apply an input logic combination which produces an incorrect output if the fault is present. It follows that the necessary and sufficient conditions for a test to detect a given fault are the following:

1. The signal value at the location of the fault should be opposite to that caused by the fault.
2. Any change of signal value at the location of the fault should cause a change of at least one output of the network.

One method of deriving such fault test patterns is through the *Boolean difference* [34]. Consider a function $f_a(x_1, x_2, \ldots, x_i, \ldots, x_n)$ which has one output value and n inputs. If one of the inputs, say, x_i, is in error, then the output would be $f_b(x_1, x_2, \ldots, \bar{x}_i, \ldots, x_n)$. To understand the action of the logic circuit when an error occurs we need to know under what circumstances the two outputs are identical. For this purpose the Boolean difference is defined as

$$\frac{df(x)}{dx_i} = f_a(x_1, x_2, \ldots, 0, \ldots, x_n) \oplus f_b(x_1, x_2, \ldots, 1, \ldots, x_n) \tag{8.24}$$

where the sign of only one parameter x_i is changed. Note that the Boolean difference is not a derivative although written as one. It simply means that if $df(x)/dx_i \equiv 0$ this implies that $f(x)$ is independent of x_i and if $df(x)/dx_i \equiv 1$ then any change in x_i will affect the output independent of the value of all x_j, $j \neq i$. In general $df(x)/dx_i$ will be a function of some (or all) the x_j's, $j \neq 1$. The set of tests for a fault on x_i is given by

$$x_i \frac{df(x)}{dx_i} = 1 \quad \text{for} \quad x_i \quad \text{s.a.0} \tag{8.25}$$

$$x_i \frac{df(x)}{dx_i} = 0 \quad \text{for} \quad x_i \quad \text{s.a.1} \tag{8.26}$$

The following operations enable us to find the Boolean difference of more complex networks:

(i) $$\frac{d\bar{f}(x)}{dx_i} = \frac{df(x)}{dx_i} = \frac{df(x)}{d\bar{x}_i}$$

(ii) $$\frac{d}{dx_i}\left(\frac{df(x)}{dx_j}\right) = \frac{d}{dx_j}\left(\frac{df(x)}{dx_i}\right)$$

(iii) $$\frac{d[f(x)g(x)]}{dx_i} = f(x)\frac{dg(x)}{dx_i} \oplus g(x)\frac{df(x)}{dx_i} \oplus \frac{df(x)}{dx_i}\frac{dg(x)}{dx_i} \tag{8.27}$$

(iv) $\quad \dfrac{d[f(x) + g(x)]}{dx_i} = \bar{f}(x)\dfrac{dg(x)}{dx_i} \oplus \bar{g}(x)\dfrac{df(x)}{dx_i} \oplus \dfrac{df(x)}{dx_i}\dfrac{dg(x)}{dx_i}$

(v) $\quad \dfrac{d[f(x) \oplus g(x)]}{dx_i} = \dfrac{df(x)}{dx_i} \oplus \dfrac{dg(x)}{dx_i}$

As an example consider the logic circuit given in Fig. 8.10 for which it is required to test for a fault at y. We can write for this circuit

$$f(x) = f = x_1 x_2 + \bar{x}_1(x_3 + x_4) = x_1 x_2 + \bar{x}_1 y$$
$$= g(x_1, x_2) + h(x_1, y) \tag{8.28}$$

where g and h are functions of x_1, x_2 and x_2, y. From Eq. (8.27(iv))

$$\dfrac{df}{dy} = \bar{g}\dfrac{dh}{dy} \oplus \bar{h}\dfrac{dg}{dy} \oplus \dfrac{dg}{dy}\dfrac{dh}{dy}$$

We note from Fig. 8.10 that $g(x)$ must be independent of y so that the functions containing dg/dy are zero. And by using the Boolean axiom $xy = \bar{x} + \bar{y}$ we can write for df/dy

$$\dfrac{df}{dy} = \overline{(\bar{x}_1 + \bar{x}_2)\bar{x}_1} \oplus 0 \oplus 0 = \bar{x}_1 \tag{8.29}$$

To test for y s.a.0 we apply Eq. (8.25) and substitute the expected logic value at y, viz.,

$$y\dfrac{df}{dy} = (x_3 + x_4)\bar{x}_1 = \bar{x}_1 x_3 + \bar{x}_1 x_4 \tag{8.30}$$

Hence by setting $x_1 = 0$, $x_3 = 1$, $x_4 = 1$ and observing the output we can see whether this is set correctly at 1 or incorrectly at 0 (s.a.0 fault condition.).
Similarly for y s.a.1

$$y\dfrac{df}{dy} = \overline{(x_3 + x_4)\bar{x}_1} = \bar{x}_1 \bar{x}_3 \bar{x}_4 \tag{8.31}$$

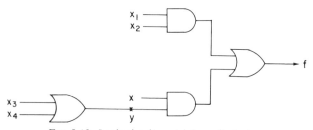

Fig. 8.10. Logic circuit containing a fault at y.

and we would need to set $x_1 = x_3 = x_4 = 0$ to detect the fault. Other similar tests are possible for simultaneous faults on two nodes in the network.

Evaluation of Eq. (8.24) is not easy to formulate for computer solution since it involves function complementation and binary addition without carry. An alternative solution based on Rademacher–Walsh techniques has been proposed by Bennetts and Hurst [11]. It is based on two theorems stated without proof.

Theorem A The spectrum of the first function shown in Eq. (8.24) is equal to the spectrum of the second function, provided that the signs of all the coefficients involving x_i are reversed. Thus taking the very simple example given in Fig. 8.1 for

$$f_a(x) = x_1 \bar{x}_2 + x_2 \bar{x}_3 \tag{8.32}$$

the spectral coefficients are

$$R_a = 0, 4, 0, -4, 4, 0, 4, 0 \tag{8.33}$$

Now if we make $x_i = x_2$ so that Eq. (8.32) becomes

$$f_b(x) = x_1 x_2 + \bar{x}_2 \bar{x}_3 \tag{8.34}$$

(sign of x_i coefficient reversed), then we obtain a new spectrum in which only the signs of R_2, $R_{1,2}$, $R_{2,3}$, and $R_{1,2,3}$ are changed thus

$$R_b = 0, 4, 0, -4, -4, 0, -4, 0 \tag{8.35}$$

Theorem B Let $f_c(x)$ and $f_d(x)$ be two functions which collectively define $f(x)$ but are mutually exclusive to $f(x)$. These are known as *disjoint functions*. Their spectra are R_c and R_d where $R_c = \text{RAD}_0^c, \text{RAD}_1^c, \ldots, \text{RAD}_{1,2}^c, \ldots, \text{RAD}_{1,2,\ldots,p}^c$, and $R_d = \text{RAD}_0^d, \text{RAD}_1^d, \ldots, \text{RAD}_{1,2}^d, \ldots, \text{RAD}_{1,2,\ldots,p}^d$ then the function

$$f_e(x) = f_c(x) \oplus f_d(x) \tag{8.36}$$

has a spectra $R_e = \text{RAD}_0^e, \text{RAD}_1^e, \ldots, \text{RAD}_{1,2}^e, \ldots, \text{RAD}_{1,2,\ldots,p}^e$, where

$$\text{RAD}_0^e = (\text{RAD}_0^c + \text{RAD}_0^d) - (m-1)2^p$$
$$\text{RAD}_1^e = (\text{RAD}_1^c + \text{RAD}_1^d)$$
$$\text{RAD}_2^e = (\text{RAD}_2^c + \text{RAD}_2^d)$$
$$\cdot$$
$$\cdot \tag{8.37}$$
$$\cdot$$
$$\text{RAD}_{1,2,\ldots,p}^e = (\text{RAD}_{1,2,\ldots,p}^c + \text{RAD}_{1,2,\ldots,p}^d)$$

That is, we simply add the spectra of the two functions term by term apart from the zero sequency coefficient from which we need to deduct a term involving m less the number of exclusive terms in $f_e(x)$.

To illustrate the formation of the Boolean difference by the use of these theorems, Theorem B is applied to the summation of the spectra given in Eqs. (8.33) and (8.35). An inverse transformation is then carried out on this summation to obtain the exclusive-OR relationship between the derived functions $f_a(x)$ and $f_b(x)$. Since $f_b(x)$ was formed from $f_a(x)$ by inverting one of its variables x_2, then this result must equal $df_a(x)/dx_2$ and express the Boolean difference of Eq. (8.24).

Thus we have

$$R_e = -8, 8, 0, -8, 0, 0, 0, 0 \tag{8.38}$$

Carrying out an inverse transformation using a matrix $\mathbf{RW}^{-1}$ (Eq. (8.3)) and dividing by $2^p = 8$ gives

$$F = -1, 1, -1, 1, -3, -1, -3, -1 \tag{8.39}$$

Replacing the negative values by a logical 1 and the positive values by a logical 0, we have

$$F(x) = 1, 0, 1, 0, 1, 1, 1, 1 \tag{8.40}$$

which will be seen to result from a truth table for $x_1 + \bar{x}_3$, which we know to be the exclusive-OR between $f_a(x)$ and $f_b(x)$, thus

$$df(x)/dx_2 = x_1 + \bar{x}_3 \tag{8.41}$$

by the definition for a Boolean difference given earlier.

Hence, what would be a considerable problem in computing the exclusive-OR operation becomes a fairly trivial operation of fast Rademacher–Walsh transformation and addition.

Further development is given by Edwards [35], who indicates how the method may be applied to the detection of single s.a.1 and s.a.0 faults in any part of the logic circuit and not necessarily at an input node. Note that this method cannot be used for multiple 'stuck-at' faults.

A similar method of logic testing is proposed by Susskind [36] to detect s.a.1 and s.a.0 faults at the primary inputs to a logic system. Its principal use would be for pin fault detection in an LSI or discrete logic system that resulted from a short circuit or open circuit at the input leads. The method determines whether a circuit has the specified input–output behaviour and not whether one or more of a specified set of faults is present as in the Boolean difference method.

Only two Rademacher–Walsh spectral coefficients are measured,

8.5 Fault Diagnosis

namely, R_0 and $R_{1,2,3,...,p}$, and used in the following manner. The binary vertices $F(x)$ for the digital circuit are derived from the truth table or an alternative method as described earlier. These binary values are then replaced by $+1$ for a logical 1 and -1 for a logical 0 and the resulting vector $\overline{F}$ is multiplied term by term with RAD 0 and RAD 1, 2, ..., p to give the spectral coefficients vectors $R_0 = \overline{F}$ RAD 0 and $R_{1,2,...,p} = R_p = \overline{F}$ RAD 1, 2, ..., p. It is shown that only these two functions are necessary to indicate whether an s.a.1 or s.a.0 fault occurs on all input leads to the system. Only the vector sum of the coefficient values for R_p and R_0 are used and these can be obtained from a simple up–down logic counter acting on the resulting values of R_p and R_0. The resulting numerical values obtained,

$$C_p = \sum_{b=0}^{b=N-1} R_p(b) \quad \text{and} \quad C_0 = \sum_{b=0}^{b=N-1} R_0(b) \qquad (8.42)$$

are used to verify the logic performance of the network. Here (b) are the coefficients of R_p, R_0 which can be either $+1$ or -1.

It may be shown that, given a function $f(x_0 x_1 \cdots x_p)$ having a $C_p \neq 0$, if in the actual implementation we measure $C_p = 0$, then one or more of the primary inputs is s.a.1. As an example of the method let

$$f(x) = x_1 x_2 + x_1 x_3 + x_2 x_3 \qquad (8.43)$$

resulting in

$$F(x) = 0, 0, 0, 1, 0, 1, 1, 1 \qquad (8.44)$$

Replacing these values with $+1$ for a logical 1 and -1 for a logical 0 gives

$$\overline{F} = -1, -1, -1, 1, -1, 1, 1, 1 \qquad (8.45)$$

From Eq. (8.1)

$$\text{RAD } 1, 2, 3 = 1, -1, -1, 1, -1, 1, 1, -1 \qquad (8.46)$$

giving

$$R_p = \overline{F} \text{ RAD } 1, 2, 3 = -1, 1, 1, 1, 1, 1, 1, -1 \qquad (8.47)$$

for which the summation of the coefficients is $C_p = 4$, i.e., $C_p \neq 0$.

It will be seen by substitution that, if one or more of x_1, x_2, x_3 are s.a.1, then the transformation by RAD 1, 2, 3 and the summation of logical values to form C_p will always be 0. A similar but slightly more complex test is applicable if $C_p = 0$. For multi-output systems a stuck-at fault on leads following the fan-out point can be detected by measuring C_0. A simple reversible-counting and comparison testing device is described by Susskind [36] for the implementation of this method.

8.5.2 Easily-tested logic networks

To complete this brief introduction to the possibilities of sequency applications in digital logic a limited discussion on the synthesis of logic networks, designed for ease of testing, will be given. The complexity of recent digital logic designs, especially VLSI implementations, raises considerable problems in adequate testing by automatic means. For production economy it is necessary to employ methods which generate a fault cover above a threshold value 95% and that fault location should be made to the smallest replaceable or repairable element. Hence there is considerable interest both in the development of suitable logical input testing sequences to reveal faults, as discussed in the preceding section, and in design methods for logic circuits which make subsequent testing easier or faster to carry out. These latter are referred to as easily-tested logic networks, and since they will almost certainly be less efficient than the minimum logic configurations that could be employed, the design criteria is to produce the best testing characteristics for the least complexity.

Several different approaches to the problem have been proposed. One method is simply to add further logic elements to a circuit to reduce the amount of testing necessary [37, 38]. A second is to design circuits specifically to make them more easily testable by using alternative (non-Boolean) design methods.

One of these, due to Reddy [39], shows that Boolean function realisations based upon Reed–Muller expansions, previously used for error-correction coding [40], can give rise to networks which are easy to test, although the circuit complexity is thereby increased. With the rapidly declining costs of semiconductor devices this may be of less importance, as has been suggested in another paper [41].

Techniques for the design of easily-tested circuits making use of various mapping and symmetry techniques and using the Rademacher–Walsh spectrum also fall into this second category [34]. It is interesting to note that there is a direct relationship between these Reed–Muller and Rademacher–Walsh methods of representing Boolean functions [42].

To a large extent the application of Rademacher–Walsh transformation to the design of easily-tested circuits involves the use of the concepts of synthesis through spectral representation and symmetries considered earlier. This is because this representation favours design with exclusive-OR circuits. The exclusive-OR function forms the basis of many easily tested networks by virtue of the result

$$\frac{df(x_i \oplus x_j)}{dx_i} = \frac{df(x_i \oplus x_j)}{dx_j} = 1 \qquad (8.48)$$

8.5 Fault Diagnosis

which establishes the conditions required for a Boolean difference approach to testing. In effect, the use of exclusive-OR gates produces easily-tested designs because every input combination to a gate is a test condition establishing a sensitive path through the gate.

A procedure is described by Edwards [35] to facilitate this translation to exclusive-OR design. The technique notes the presence of a high-valued secondary coefficient in the spectrum for the logic function $f(x)$ and uses this to identify an exclusive-OR sub-circuit in the actual implementation. Extracting this coefficient from R_i leads to a modified form of the spectrum R'_i, which defines the remaining function to be implemented. A further high-valued secondary coefficient in R'_i is used to identify another exclusive-OR sub-circuit to be incorporated into the final synthesis, and the process is repeated until only the primary coefficients remain.

There is a further advantage of the spectral representation of logical functions: this arises from the observation that the functions which are easiest to synthesise are those which have their highest-magnitude components in the first order positions $0, 1, \ldots, p$. We can see this from the correlation relationship pointed out in Subsection 8.2.1. If the spectrum has a large component at, say r_2, then this is associated with x_2 and could be synthesised by simply connecting the input x_2 to the output. On the other hand, a large component situated at $r_{1,2}$ would require an exclusive-OR implementation with consequent increase in circuit complexity. Hence there is considerable advantage in selecting logic circuits in which the spectral representation is concentrated towards the first $p + 1$ spectral positions. To aid this selection a spectral translation procedure can be applied which moves certain high-order spectral components into low-order positions.

The following theorem for this translation is proved by Edwards [19, p. 53]: 'If in a Boolean function $f(x_1, \ldots, x_k, \ldots, x_n)$ having a spectrum R_i, x_k is replaced by $[(x_a \oplus x_b, \ldots, \oplus x_h) \oplus x_k]$, then the new spectrum R'_i so formed can be generated from R_i by deleting the subscript k where it appears in the subscript for $R_i (i = 0, 1, 2, \ldots, 12, \ldots, 12 \ldots p)$ and adding it where it does not appear.' Thus if we take a function $f(x_1 x_2 x_3)$ and replace x_1 by $(x_1 \oplus x_2)$, then

$R'_{1,2} = R_1,$	$R'_{1,2,3} = R_{1,3}$	(subscripts deleted)
$R'_1 = R_{1,2},$	$R'_{1,3} = R_{1,2,3}$	(subscripts added)
$R'_{2,3} = R_{2,3},$	$R'_2 = R_2,\quad R'_3 = R_3$	(subscripts unchanged)

The effect this has on the ordering of the spectral coefficients can be seen if we take a practical example. Consider the function

$$f(x) = (x_1 + \bar{x}_2 + x_3)(x_1 + x_2 + \bar{x}_3) \tag{8.49}$$

shown in Fig. 8.11a in terms of AND/OR/NOT elements. From its truth table $F(x)$ is derived as 11011011 for which the Rademacher–Walsh spectrum is

$$R_i = -4, 0, 0, 0, -4, 4, -4, 0 \qquad (8.50)$$

By replacing x_1 by $(x_1 \oplus x_2)$ a new spectrum is generated through the theorem to give

$$R'_i = -4, -4, 0, 0, 0, 0, -4, 4 \qquad (8.51)$$

Repeating this operation for x_2 replaced by $(x_2 \oplus x_3)$ gives

$$R''_i = -4, -4, -4, 0, 4, 0, 0, 0 \qquad (8.52)$$

The spectral terms will be seen to be shuffled down towards the primary coefficient end and point clearly to an alternative statement of the function given in Eq. (8.49) as

$$f''(x''_1, x''_2, x''_3) = (\bar{x}''_1 + \bar{x}''_2) \qquad (8.53)$$

where $x''_1 = x_1 \oplus x_2$ and $x''_2 = x_2 \oplus x_3$.

This is given in terms of the logical addition of the outputs of two exclusive-OR gates represented by $\bar{x}''_1$ and $\bar{x}''_2$ in Fig. 8.11b. Another theorem of

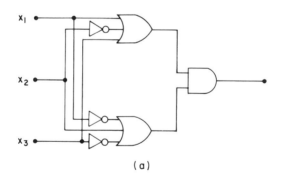

(a)

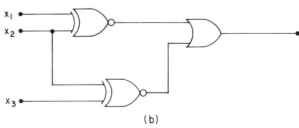

(b)

FIG. 8.11. Alternative logic circuits: (a) $f(x) = (x_1 + \bar{x}_2 + x_3)(\bar{x}_1 + x_2 + \bar{x}_3)$; (b) $f(x) = (x_1 \oplus x_2) + (x_2 \oplus x_3)$.

disjoint spectral translation enables several spectral components to be translated simultaneously to lower orders [19, 35].

All of these techniques of synthesis through spectral representation, use of symmetries, spectral translation, disjoint spectral translation and generation of test sequences through Boolean differences assist in the design of logical circuits that are easy to test. They form the basis of a new set of development techniques which are currently being worked out for the design and fabrication of large-scale integrated circuit systems [43]. These include recent developments in the detection of transient errors generated within the logic module, often referred to as 'soft errors'. These have been analysed in complex VLSI logic by Redinbo [44], who postulates the use of Walsh transform domain quantities in a probabilistic analysis of error for both single- or multiple-output functions.

References

1. Mano, M. M. (1972). "Computer Logic Design." Prentice-Hall, Englewood Cliffs, New Jersey.
2. McClusky, E. J. (1962). Minimal sums for Boolean functions having many unspecified products. *IRE J. Circ. Syst.* **CT-9**, 387–392.
3. Roth, J. P. (1959). Algebraic topological methods in synthesis. *Proc. Int. Symp. Theory Switching,* pp. 57–73. Harvard Univ. Press, Cambridge, Massachusetts.
4. Coleman, R. P. (1961). Orthogonal functions for the design of switching circuits. *IEEE Trans. Electron. Comput.* **EC-10**, 379–383.
5. Lechner, R. J. (1970). A transform approach to logic design. *IEEE Trans. Comput.* **C-19**, 627–640.
6. Karpovsky, M. G. (1976). "Finite Orthogonal Series in the Design of Digital Devices." Wiley, New York.
7. Hurst, S. L. (1978). "Logical Processing of Digital Signals." Crane Russak, New York; Edward Arnold, London.
8. Dertouzos, M. L. (1965). "Threshold Logic: A Synthesis Approach. MIT Res. Monograph 32. MIT Press, Cambridge, Massachusetts.
9. Chow, C. K. (1961). On the characterisation of threshold functions. *Proc. IEEE Symp. Switching Theory Logic Design,* IEEE Special Publication S134, pp. 34–38.
10. Edwards, C. R. (1978). A special class of universal logic gates and their evaluation under a Walsh transform. *Int. J. Electron.* **44**, 49–59.
11. Bennetts, R. G., and Hurst, S. L. (1978). Rademacher–Walsh spectral transform: A new tool for problems in digital network fault diagnosis? *IEE J. Comput. Dig. Tech.* **1**, 38–44.
12. Lloyd, A. M. (1979). A consideration of orthogonal matrices, other than the Rademacher–Walsh types, for the synthesis of digital networks. *Int. J. Electron.* **47**, 205–212.
13. Henderson, K. W. (1964). Some notes on the Walsh functions. *IEEE Trans. Electron. Comput.* **EC-13**, 50–52.
14. D'Alton, L. B. (1980). Rademacher–Walsh diagnosis. *Electron. Eng. (Great Britain)* **52** (643), 83–87.
15. Muzio, J., Miller, D. M., and Hurst, S. L. (1982). Number of spectral coefficients necessary to identify a class of Boolean functions. *Electron. Lett.* **18**, 577–578.

16. Edwards, C. R. (1973). The application of the Rademacher–Walsh transform to digital circuit synthesis. *Proc. Theory Applic. Walsh Functions, Hatfield Polytechnic, England.*
17. Hurst, S. L. (1971). "Threshold Logic." Pitman Press, Bath, England.
18. Winder, R. O. (1964). Threshold functions through $n = 7$. *RCA Lab. Sci. Report 7*, Princeton, New Jersey.
19. Edwards, C. R. (1975). The application of the Rademacher–Walsh transform to Boolean function classification and threshold logic synthesis. *IEEE Trans. Comput.* **C-24**, 48–62.
20. Golomb, S. W. (1959). On the classification of Boolean functions. *IRE Trans. Comput.* **CT-6**, 176–186.
21. Bennett, L. A. M. (1982). Realisation of logic functions by multi-output threshold logic gates. *IEE Proc. E* **129**, 239–243.
22. Picton, P. D. (1981). Realisation of multi-threshold logic networks using the Rademacher–Walsh transform. *IEE Proc. E* **128**, 107–113.
23. Hurst, S. L. (1977). Detection of symmetries in combinatorial functions by spectral means. *IEEE J. Electron. Circ. Syst.* **1**, 173–180.
24. Lloyd, A. M. (1978). Spectral addition techniques for the synthesis of multivariable logic networks. *IEE J. Comput. Dig. Tech.* **1**, 152–163.
25. Muzio, J. C., Miller, D. M., and Hurst, S. L. (1984). "Spectral Techniques in Digital Logic Design." Academic Press, London.
26. Lloyd, A. M., and Hurst, S. L. (1979). Spectral summation techniques for the synthesis of digital logic networks. *Proc. IEE Int. Conf. CAD Manuf. Electron. Comput. Syst., Sussex, England*, pp. 66–70.
27. Besslich, P. W. (1978). On the Walsh–Hadamard transform and prime implement extraction. *IEEE Trans. Electromag. Compat.* **EMC-20**, 516–519.
28. Besslich, P. W. (1979). Computer-aided design of logic circuits using transform methods. *Proc. IEE Int. Conf. CAD Manuf. Electron. Comput. Syst., Sussex, England*, pp. 75–79.
29. Besslich, P. W. (1978). Determination of the irredundant forms of a Boolean function using Walsh–Hadamard analysis and dyadic groups. *IEE J. Comput. Dig. Tech.* **1**, 143–151.
30. Besslich, P. W. (1981). Fast transform procedure for the generation of near-minimal covers of Boolean functions. *IEE Proc. E.* **128**, 250–254.
31. Dunham, B., and Fridshol, R. (1959). The problem of simplifying logical expressions. *J. Symbol. Logic* **24**, 17–19.
32. Besslich, P. W. (1983). A method for generation and processing of dyadic indexed data. *IEEE Trans. Comput.* **C-32**, 487–493.
33. Hurst, S. L. (1974). The application of Rademacher–Walsh spectra to the testing of logic networks. *Proc. Conf. Automat. Testing, Brighton, England*, pp. 15–25.
34. Sellars, F. F., Hsiang, M. Y., and Bearnson, L. W. (1968). Analysing errors with the Boolean difference. *IEEE Trans. Comput.* **C-17**, 676–683.
35. Edwards, C. R. (1977). The design of easily tested circuits using mapping and spectral techniques. *Radio Electron. Eng.* **47**, 321–342.
36. Susskind, A. K. (1981). Testing and verifying Walsh coefficients. *IEEE Ann. Symp. Fault-Tolerant Computing, 11th, Portland, Oregon*, pp. 206–208.
37. Hayes, J. P. (1974). On modifying logic networks to improve their diagnosability. *IEEE Trans. Comput.* **C-23**, 56–62.
38. Saluja, K. K., and Reddy, S. M. (1974). On minimally testable logic networks. *IEEE Trans. Comput.* **C-23**, 552–554.
39. Reddy, S. M. (1972). Easily testable realisations for logic functions. *IEEE Trans. Comput.* **C-21**, 1183–1188.
40. Reed, I. S. (1954). A class of multiple-error-correcting codes and the decoding scheme. *IRE Trans. Info. Theory* **PGIT-4**, 38–49.

41. Page, E. W. (1980). Minimally testable Reed–Muller canonical forms. *IEEE Trans. Comput.* **C-29,** 746–750.
42. Stanković, R. S. (1982). A note of the relation between Reed–Muller expansions and Walsh transforms. *IEEE Trans. Electromag. Compat.* **EMC-24,** 68–70.
43. Symposium (1983). *Int. Workshop Fault Detection Spectral Tech., Boston, October 12–14.*
44. Redinbo, G. R., and Wang, G. S. (1983). Probability of error in combinational logic systems containing soft faults. *IEE Proc. E.* **130,** 125–137.

Selected List of Additional References

Mathematical theory

1. Bansal, V. S. (1978). A class of describing functions for dyadic systems. *J. Franklin Inst.* **306,** 275–281.
2. Bansal, V. S., and Mirza, K. B. (1978). On a class of describing functions for dyadic systems. *J. Franklin Inst.* **306,** 275–281.
3. Burkhardt, H. (1980). On invariant sets of a certain class of fast translation invariant transforms. *IEEE Trans. Acoust., Sp. Sig. Proc.* **ASSP-28,** 517–523.
4. Butzer, P. L., and Splettstösser, W. (1978). Sampling principle for duration-limited signals and dyadic Walsh analysis. *Inf. Sci.* **14,** 93–106.
5. Chen, C. F. (1977). Walsh operational matrices for fractional calculus and their application to distributed systems. *J. Franklin Inst.* **303,** 267–284.
6. Cheng, D. K. (1976). Time-shift theorems for Walsh transforms and a solution to difference equations. *IEEE Trans. Electromag. Compat.* **EMC-18,** 83–87.
7. Engels, W., and Splettstösser, W. (1982). On Walsh differentiable dyadically stationary random processes. *IEEE Trans. Info. Theory* **IT-28,** 612–619.
8. Kunz, H. O. (1979). On the equivalence between one-dimensional discrete Walsh-Hadamard and multi-dimensional discrete Fourier transforms. *IEEE Trans. Comput.* **C-28,** 267–268.
9. Lilein, A. L. (1981). The Walsh spectrum transformation when there is a signal shift. *Telecom. Radio Eng. 2 (USA)* **36,** 50–55.
10. Morettin, P. A. (1976). Estimation of the Walsh spectrum. *IEEE Trans. Info. Theory* **IT-22,** 106–107.
11. Moricz, F. (1981). On Walsh series with coefficients tending monotomically to zero. *Acta Math. H.* **38,** 183–189.
12. Nagai, T. (1976). Dyadic stationary processes and their spectral representations. *Bull. Math. Statist.* **17,** 65–73.
13. Onneweer, C. W. (1976). Convergence of Walsh functions. *North Am. Math.* **A679,** 23–27.

14. Retter, M. L. (1975). Structural approach to modern integral transform theory. *Int. Symp. Walsh Function Applic. Hatfield Polytechnic, England.*
15. Richards, J. (1979). Anharmonic analysis of lattice field theories. *Phys. Rev. D* **20**, 1351–1359.
16. Splettstösser, W. (1980). Error analysis in the Walsh sampling theorem. *IEEE Symp. Electromag. Compat., Baltimore*, pp. 366–370.
17. Splettstösser, W., and Engels, W. (1982). A note on truncation error bounds in the dyadic (Walsh) sampling theorem. *IEEE Trans. Acoust., Sp. and Sig. Proc.* **ASSP-30**, 2.
18. Van Till, J. W. J. (1973). Comments on the definition and generation of Walsh functions. *IEEE Trans. Comput.* **C-22**, 702–703.
19. Waterman, D. (1982). On systems of functions resembling the Walsh system. *Mich. Math. J.* **29**, 83–87.
20. Yuen, C. K. (1975). Function approximation by Walsh series. *IEEE Trans. Comput.* **C-24**, 590–598.

Transformation

1. Algazi, V. R. (1982). Performance and computation ranking of fast unitary transforms in applications. *IEEE Cong. Acoust., Sp. and Sig. Proc., Paris*, pp. 32–35.
2. Besslich, P. W. (1982). Transform pre-processing of binary patterns for structural classification. *Proc. Int. Conf. Pattern Recognition, 6th, Munich*, pp. 331–334.
3. Cheng, D. K., and Lin, J. J. (1977). An algorithm for sequency ordering of Hadamard functions. *IEEE Trans. Comput.* **C-26**, 308–309.
4. Fino, B. J., and Algazi, V. R. (1976). Unified matrix treatment of the fast Walsh–Hadamard transform. *IEEE Trans. Comput.* **C-25**, 1142–1146.
5. Hama, H., and Yamashita, K. (1979). Walsh–Hadamard power spectra invariant to certain transform groups. *IEEE Trans. Syst., Man Cybernet.* **SMC-9**, 227–237.
6. Hostetter, G. H. (1984). Recursive discrete computation of Walsh–Hadamard transformations. Signal Processing, North Holland [to be published].
7. Rao, K. R., and Ahmed, N. (1971). Complex BIFORE transform. *Int. J. Syst. Sci.* **2**, 149–162.
8. Slutter, J. A. (1982). Generalized running discrete transform. *IEEE Trans. Acoust., Sp. Sig. Proc.* **ASSP-30**, 60–68.
9. Trachtenberg, E. A. (1980). Construction of fast unitary transforms which are equivalent to Karhunen-Loève spectral representations. *IEEE Symp. Electromag. Compat., Baltimore*, pp. 376–379.
10. Yip, P.C. (1976). Zoom Walsh transform. *IEEE Trans. Electromag. Compat.* **EMC-18**, 79–83.
11. Yuen, C. K. (1982). On the Walsh transform of a shifted vector. Univ. of Hong Kong. Publ. TR-10-82, Hong Kong.

Signal processing

1. Briscoe, W. L., and Shipley, J. P. (1976). Some comparisons of Walsh transformations and Fourier transformations for speech compression. Los Alamos Lab., NTIS Report LA-6512-MS, Los Alamos, New Mexico.
2. Gulamhusein, M. N., and Fallside, F. (1973). Shot-time spectral and autocorrelation analysis in the Walsh domain. *IEEE Trans. Info. Theory* **IT-19**, 615–623.

3. Hermann, G. Y. (1982). Real-time monitoring of machine tools via Walsh–Hadamard transform. *IEEE Conf. Acoust., Sp. Sig. Proc.,* pp. 343–346.
4. Herron, R. L. (1977). Comparison of fast Fourier transforms and other transforms in signal processing for tactical radar target identification. Air Force Inst. of Tech., NTIS Report AD-AO53 347/1ST, Wright-Patterson Air Force Base, Ohio.
5. Hutchinson, B. A. (1978). Small business telephone systems. Ph.D. thesis, Univ. of Newcastle, Newcastle, England.
6. Lyle, W. D., and Forte, F. (1980). A useful property of the coefficients of a Walsh–Hadamard transform. *IEEE Trans. Acoust., Sp. Sig. Proc.* **ASSP-28,** 479–480.
7. Maqusi, M. (1977). Walsh analysis of power-law devices. *IEEE Trans. Info. Theory* **IT-23,** 144–146.
8. Maqusi, M. (1982). Correlation and spectral analysis of non-linear transformations of sequency band-limited signals. *IEEE Trans. Acoust., Sp. Sig. Proc.* **ASSP-30,** 513–516.
9. Reddy, N. S. (1981). New algorithms for fast digital convolution: A review. *Electro-Technol. (India)* **25,** 176–182.
10. Thaker, G. H. (1977). Comparison of fast Fourier and fast Walsh transform methods in speech recognition. *IEEE Proc. Southeast Conf.,* pp. 419–432.
11. Yip, P., and Hutchinson, D. (1982). Residual correlation for generalized discrete transforms. *IEEE Trans. Electromag. Compat.* **EMC-24,** 64–68.

Image processing

1. Andrews, H. C., and Hunt, B.R. (1977). "Digital Image Restoration." Prentice-Hall, Englewood Cliffs, New Jersey.
2. Besslich, W. (1980). A W.H.T. method for generation and processing of subsumed and average data. *IEEE Symp. Electromag. Compat., Baltimore,* pp. 302–306.
3. Hama, H. (1978). On the theory of feature extraction for geometrical patterns, *in* "Progress in Cybernetics and Systems Research" (R. Trappl, ed.), pp. 473–483. Wiley, New York.
4. Lux, P. (1977). A novel set of closed orthogonal functions for picture coding. *Arch. Elektronik Übertragungstechnik* **31,** 267–274.
5. Lynch, R.T., and Reis, J. J. (1976). Haar transform image coding. *Nat. Telecom. Conf., Dallas,* pp. 443–447.
6. Mount, F. W., Netravali, A. N., and Prasada, B. (1977). Design of quantizers for real-time Hadamard transform coding of pictures. *Bell Syst. Tech. J.* **56,** 21–48.
7. Netravali, A. N., Prasada, B., and Mounts, F. W. (1977). Some experiments in adaptive and predictive Hadamard transform coding of pictures. *B.S.T.J.,* **56,** 1531–1547.
8. Ohta, T. (1980). Rademacher transform image coding. *IEEE Symp. Electromag. Compat., Baltimore,* pp. 282–287.
9. Yip, P., and Rao, K.R. (1978). Sparse-matrix factorization of discrete sine transform. *Proc. Ann. Asilomar Conf. Circ. Syst. Comput., 12th, Pacific Grove, California,* pp. 549–555.

Communication

1. Bourbakis, N. (1982). An efficient real-time method for transmitting Walsh–Hadamard transformed pictures. *IEEE Conf. Acoust., Sp. Sig. Proc.,* **ASSP-28,** 452–455.
2. Chaiko, K. I. (1980). Use of Walsh functions for data transmission over channels with short-term interruption. *Telecom. Radio Eng. 2 (USA)* **35,** 79–83.

3. Chen, W. L. (1982). Walsh series analysis of multi-delay systems. *J. Franklin Inst.* **313**, 207–217.
4. Handelsman, M. (1972). Time domain impulse antenna study. Rome Air Develop. Center Report TR-72-105, NTIS Report AD744837, Washington, D.C.
5. Knab, J. J. (1977). Effects of round-off noise on Hadamard transformed imagery. *IEEE Trans. Commun.* **COM-25**, 1292–1294.
6. Leong, S. H. (1981). Radar resolution based on non-sinusoidal waveforms. Ph.D. thesis, Catholic Univ. of America, Washington, D.C.
7. Lin Zhang-Kan and Zhang Qisham (1980). Walsh cross-correlation functions and choice of Walsh sub-carriers. *J. Beijung Inst. Aeronauts.* **1**, 73–81.
8. Schlichta, P. J. (1979). Higher dimensional Hadamard matrices. *IEEE Trans. Info. Theory* **IT-25**, 566–572.
9. Scrinivasan, V. K., and Vantron, C. (1982). On an approximation theorem of Walsh in the radio field. *J. Approx. Theory* **35**, 191–193.
10. Susman, L., and Lamensdorf, L. (1980). Picosecond pulse antenna techniques. Rome Air Develop. Center Report TR-71-64, NTIS Report AD884646, Washington, D.C.
11. Thompson, K. R. (1977). Analysing a bi-orthogonal information channel by the Walsh–Hadamard transform. *Comp. Electron. Eng. (USA)* **4**, 119–132.
12. Van Cleave, J. (1980). Walsh pre-processor. Am. Electron. Lab., NTIS Report AD-AO91, 188/3, Lonsdale, Pennsylvania.
13. Wong, K. M., and Jan, Y. G. (1983). Adaptive Walsh equaliser for data transmission. *IEE Proc. F* **130**, 153–160.
14. Zheng, H., and Youmei, Y. (1980). An adaptive multiplex delta-modulation system. *IEEE Symp. Electromag. Compat., Baltimore*, pp. 346–348.

Systems and control

1. Bailey, J. (1980). Application of Walsh functions to linear systems. *IEEE Proc. Southeastern Conf.*, pp. 58–59.
2. Bohn, E. V. (1982). Measurement of continuous time linear system parameters via Walsh functions. *IEEE Trans. Ind. Eng.* **29**, 38–46.
3. Cameron, R. (1980). A new approach to the prediction of limit cycles. *Int. J. Control.* **32**, 963–981.
4. Chen, C. F. (1976). Design of piece-wise gains for optimum control via Walsh functions. *IEEE Trans. Automat. Control* **21**, 635–636.
5. Chen, W. L., and Lee, C. L. (1982). Walsh series expansion of composite functions and application to linear systems. *Int. J. Systems* **13**, 219–226.
6. Corrington, M. S. (1973). Solution of differential and integral functions with Walsh functions. *IEEE Trans. Circuit Theory* **CT-20**, 470–476.
7. Karanan, V. R. (1978). Bi-linear system identification by Walsh functions. *IEEE Trans. Automat. Control* **AC-23**, 709–713.
8. Kaul, L. K. (1977). Evaluation of non-linearities using Walsh input. *J. Inst. Electron. Telecom (India)* **22**, 767–771.
9. McCarthy, E. P. (1978). Novel method of direct digital integrated cycle power control. *J. Franklin Inst.* **306**, 267–274.
10. Palanisamy, K. R. (1981). Analysis and optimum control of linear systems via single term Walsh series approach. *Int. J. Systems Sci.* **12**, 443–454.
11. Palanisamy, K. R. (1982). Analysis of non-linear system vai single term Walsh function approach. *Int. J. Systems Sci.* **13**, 929–935.

12. Rao, G. P., and Sivankumer, L. (1981). Transfer function matrix identification in MIMO system via Walsh functions. *Proc. IEEE* **69**, 465–466.
13. Rao, G. P. (1982). Order and parameter identification in continuous linear systems via Walsh functions. *IEEE Proc.* **70**, 764–766.
14. Rao, G. P. (1982). Piece-wise linear system identification via Walsh functions. *Int. J. Systems Sci.* **13**, 525–530.
15. Rao, G. P., and Srinivasan, T. (1980). Multidimensional block-pulse functions and their use in the study of distributed parameter systems. *Int. J. Systems Sci.* **11**, 689–708.
16. Sinha, M. S. P., Rajamari, V. S., and Sinha, A. K. (1980). Identification of non-linear distributed system using Walsh functions. *Int. J. Control* **32**, 669–676.
17. Subbayyan, R. (1979). Walsh functions approach for simplification of linear systems. *IEEE Proc.* **67**, 1676–1678.
18. Tzafestas, S. (1977). Nuclear reactor control using Walsh function variational synthesis. *Nucl. Sci. Eng.* **62**, 763–770.
19. Tzafestas, S. (1978). Walsh series approach to lumped and distributed system identification. *J. Franklin Inst.* **305**, 199–220.

Optical applications

1. Decker, J. A. (1972). Hadamard transform spectrometry: A new analytical technique. *Anal. Chem.* **44**, 127.
2. Despain, A. M., and Vanasse, G. A. (1972). Walsh functions in spectroscopy. *Proc. Symp. Applic. Walsh Functions, Washington, D.C.*, AD744650, pp. 30–35.
3. Fukui, I. (1982). Two-dimensional Walsh transforming device using an integrated optical circuit. *IEEE Trans. Circ. Syst.* **CAS-29**, 336–339.
4. Harwit, M. (1971). Spectrometric images. *Appl. Opt.* **10**, 1415–1421.
5. Harwit, M., and Sloan, J. A. (1979). "Hadamard Transform Optics." Academic Press, New York.
6. Hazra, L. N. (1977). A new class of optimum amptitude filters. *Optics Comm.* **21**, 232–236.
7. Hazra, L. N. (1978). Walsh functions in lens optimization. *Optica Acta* **25**, 573–584.
8. Hazra, L. N. (1979). Apodization of aberrated pupils. *Can. J. Phys.* **57**, 1340–1346.
9. Purkait, P. K. (1981). Walsh function in lens optimization. *Optica Acta* **28**, 389–396.

Logical analysis

1. Besslich, P. W. (1983). Spectral processing using WHT-related in-place transformation. *Workshop Fault Detection Spectral Techn., Boston.*
2. Chen, X., and Hurst, S. L. (1982). A comparison of universal logic module realization and their application in the synthesis of combinational and sequential logic networks. *IEEE Trans. Comput.* **C-31**, 140–147.
3. Cheng, D. K. (1979). Stochastic behavior of digital combinational circuits. NTIS Report AD-AO68, 754/1ST, Washington, D.C.
4. Edwards, C. R., and Hurst, S. L. (1978). A digital synthesis procedure under function symmetries and mapping methods. *IEEE Trans. Comput.* **C-27**, 985–997.
5. Hurst, S. L. (1979). An engineering consideration of spectral transforms for ternary logic synthesis. *Computer J.* **22**, 173–183.
6. Hurst, S. L. (1980). Custom LSI design: The universal logic module approach. *IEEE Proc. Int. Conf. Circ. Comput. New York*, pp. 1116–1119.

7. Hurst, S. L. (1981). The Haar transform in digital network synthesis. *IEEE Proc. Int. Symp. Multiple Valued logic, 11th,* pp. 10–18.
8. Karpovsky, M. (1981). An approach for error detection and error correction in distributed systems computing numerical functions. *IEEE Trans. Comput.* **C-30,** 947–953.
9. Karpovsky, M. (1982). Universal tests detecting input/output faults in almost all devices. *IEEE Int. Test Conf.,* pp. 52–57.
10. Karpovsky, M., and Leviton, L. (1982). Detection and identification of stuck-at faults in combinational and sequential VLSI networks by universal tests. State Univ. of New York, Tech. Report CS82-03, Binghampton, New York.
11. Muzio, J. C., Miller, D. M., and Hurst, S. L. (1984). "Spectral Techniques in Digital Logic Design." Academic Press, London.
12. Shanker, A. U., and Cheng, D. K. (1979). Noise error determination of combinational circuits by Walsh functions. *IEEE Trans. Electromag. Compat.* **EMC-21,** 146–152.

Symposia

A number of symposia on the applications of Walsh functions were held annually in Washington, D.C., until they were incorporated into the annual symposia of the IEEE Electromagnetic Compatability Group. They now take place in various United States locations and are published as regular IEEE proceedings. These earlier Washington proceedings are available from the National Technical Information Service, U.S. Department of Commerce, Springfield, Virginia 22151:

1970 (C. A. Bass, ed.), AD707431
1971 (R. W. Zeek and A. E. Showalter, eds.), AS727000
1972 (R. W. Zeek and A. E. Showalter, eds.), AD744650
1973 (R. W. Zeek and A. E. Showalter, eds.), AD763000
1974 (H. Schreiber and G. P. Sandy, eds.) [available from the IEEE, New York, under order no. 74CH0861-5EMC]

Bibliographies

1. Bramhall, J. N. (1974). An annotated bibliography on Walsh and Walsh-related functions. John Hopkins Univ. Applied Physics Lab., Tech. Memoir TG 1198B, Baltimore, Maryland.
2. National Technical Information Service Data Base, Walsh Functions (1964–1982), PB83-800938. U.S. National Technical Information Service, Springfield, Virginia.

Index

A

Adaptive majority multiplexer, 240
  threshold scheme, 128
Added-pulse generation, 140
Addition
  modulo-2, 28, 51, 100, 269
  relationship, 53
  theorem, 100
Aerial, *see* Radiator
Aliasing, 44, 154
Algorithm
  Berauer's, 149
  bit reversal, 20, 71, 157
  constant geometry, 72, 162
  Cooley–Tukey, 20, 65
  fast transform, 49
  Fourier transform, 99
  in-place, 63, 71
  Walsh transform, 49
Amplifier
  sample and hold, 122, 149
  sum/differing, 149
Analysis, 98
Arithmetic autocorrelation, 108
Array generators, 138
  logic, 159
  two-dimensional, 93

Associative property, 160
Autocorrelation, 98, 251, *see also* Correlation
  arithmetic, 108
  dyadic, 108
  function, 108
Automatic coding system, 245
Averaged Walsh power spectrum, 110

B

Band limiting, 233
Beam forming, 260
  pattern, 259
Berauer's algorithm, 149
Besslich Rademacher–Walsh matrix, 279
Binary–Walsh matrix, 266
Bit inversion, 71
Bit reversal, 20, 157
Block pulse carrier, 233, 251
Block pulse multiplexer, 233
Boolean difference, 283
  functions, 264
  logic analysis, 265
  logic expression, 161
  synthesis, 24, 30
  truth table, 269

Butterfly, 62, 145, 212
 cosine/sine, 207
 diagram, 63
 reversed, 63

C

CAL function, 15, 22, 50
CAL-SAL ordering, 21, 74
CAL-SAL transformation, 74
Carrier-free radar, 253
Cepstrum, 129
Channel capacity, 245
Channel noise, 238
Charged-coupled device, 161
Chirp-Z transform, 162
Chow parameter, 265, 268
Circular time shift, 51, 104
Closed series, 4
C matrix transform, 206
Code alphabet, 240
 mean-square-error, 245
 morse, 115
Coding, 240
 automatic, 245
 error correction, 288
 group, 244
 Reed-Muller, 244
 variable word length, 244
Communication, 231
Compensation matrices, 91
Complex Fourier transform, 207
Compressional efficiency, 206
Computer-aided design, 161
Constant-geometry algorithm, 162
 flow diagram, 147
 structure, 73
 transform, 150
Continued product definition, 30
Continued product representation, 59
Contrast enhancement, 215
Convergence, 40, 132
Conversion, 55
 from matrix, 63
Convolution, 99
Cooley-Tukey algorithm, 20, 65
Correlation, 98, 202
 arithmetic, 108
 auto-, 98, 251
 cross-, 101

Cosine function, 5
Cosine/sine butterflies, 207
Cosine transform coefficients, 111
Cospec function, 207
Cross-correlation, 101, *see also* Correlation
Cross-talk, 236
C-T, *see* Cooley-Tukey algorithm
Current-loop dipole, 255

D

Data compression, 211
Decimation in frequency, 65
Decision theoretic approach, 220
Degree order, 38
Derivation of Walsh series, 24
Detection of symmetries, 271
Diagonal matrix, 91
Diagram method, 264
Difference method, 25
Differential pulse-code modulation, 213
Digital filtering, 118
 logic, 140
 sequency multiplexing, 232, 236
Dipole
 electric, 255
 folded, 256
 magnetic, 255
 radiation, 257
 stacked magnetic, 256
Direct memory access, 157
Discrete cosine transform, 206
Discrete Fourier transform, 51, 89
Discrete Haar transform, 78
Discrete sampled functions, 44
Discrete slant transform, 82
Discrete Walsh function, 11
Discrete Walsh transform, 49
Disjoint functions, 285
Disjoint spectral translation, 270, 291
DMA, *see* Direct memory access
Dominant-term concept, 130
Doppler effect, 248
Dyadic autocorrelation, 108
Dyadic convolution, 100, 129
Dyadic domain, 102
Dyadic order, 20
Dyadic time shift, 51, 102
Dyadic translation, 51

E

Easily tested logic, 288
Edge detection, 214
Eigen values, 48
Electric dipole, 255
Electromagnetic radiation, 246
Electromagnetic Walsh radiation, 247
Equivalence symmetry, 272
Error codes, 245
Error control, 242
Error correlation, unequal, 245
Error correlation coding, 288
Error orthogonality, 140
Exclusive-OR gate, 266, 269, 289

F

False minterm, 276
Fast cosine transform, 206
Fast discrete transform, 49
Fast Fourier transform, 51, 65, 87, 89
Fast Haar transform, 78
Fast slant transform, 84
Fast transform algorithms, 49
Fast Walsh transform, 58
Fault diagnosis, 281
Feature extraction, 222
FFT, *see* Fast Fourier transform
FFT-derived transforms, 69
Filtering, 118
Filter
  Gaussian, 218
  inverse, 219
  matched, 123, 218
  matrix, 120
  non-linear, 129
  optimal, 218
  scalar, 120
  sequency, 215
  threshold, 122, 223
  transversal, 161
  two-dimensional, 127
  vector, 120
  Walsh, 104, 124
  weights, 123
  Wiener, 119
Flow diagram, 61
  constant-geometry, 147

Fourier analysis, 48
  from Walsh transformation, 90
Fourier coefficient, 7
Fourier filtering, 124
Fourier periodogram, 110
Fourier series, 17, 36
Fourier transform, 51, 89, 99
Fourier transform algorithm, 20
Frequency
  division, 232
  limited power spectrum, 103
  limited signal, 90
Functions, 4
  CAL, 15, 22, 50
  CAL-SAL, 21
  cosine, 5, 17
  cospec, 207
  Haar, 38
  logic, 271
  orthogonal, 5
  polynomial, 14
  Rademacher, 8, 20, 27, 138, 266
  SAL, 15, 22, 50
  sampled, 44
  sequency, 15
  sine, 5, 17
  slant, 13, 43, 82
  symmetric, 87
  time-limited, 103
  two-dimensional, 202
  Walsh, 11
Function ordering, *see* Ordering
FWT, *see* Fast Walsh transform

G

Gaussian filtering, 218
  form, 203
  noise, 127
Generator
  global, 144, 145
  programmable, 144
  serial programmable, 143
  single-chip Walsh, 158
  Walsh, 138
Global generator, 144, 145
  transform, 216
Gradient detection, 215
Gray code, 28, 245
  conversion, 145, 157

Gray level, 14
Group codes, 244

H

Haar functions, 11, 132
Haar function matrix, 78
Haar function series, 38
  two-dimensional, 217
Haar-like series, 217
Haar matrix, 78
Haar power spectrum, 112, 222
Haar slant series, 43, 85
Haar transform, 75
Hadamard matrix, 24, 28, 30, 166
Hadamard order, 20, 39
Hadamard transform, 205, 279
Hamming distance, 242
Hardware function generation, 31
Hardware techniques, 137
Harmonic motion, 114
Harmonic number, 19
Harmonic spectrum, 103
Harmuth array generator, 139
Harmuth phasing, 23
Hybrid series, 40
Hybrid transformation, 82

I

Ideograms, 226
Image coding, 206
Image compression, 201, 204, 211
Image enhancement, 126, 202, 214
Image matrix, 202
Image processing, 201
Image restoration, 214, 219
Image transformation, 204
Image transmission, 210
Impulse radar, 247, 251, 253
Incomplete function set, 7
Incompletely specified logic, 278
Incomplete orthogonal function series, 5
Incremental lag, 99
In-place algorithm, 63, 71
Intel microprocessor, 154
Inter-frame compression, 212
Interstage shuffling, 148

Into-the-ground radar, 253
Intra-frame colour television, 211
Inverse filtering, 219

K

Kanji characters, 226
Karhunen–Loève series, 14
Karhunen–Loève transform, 48, 83
Karnaugh map, 264
Kronecker multiplication, 66, 83
Kronecker products, 20, 43, 58

L

Large-scale integration, 158
Linguistic approach, 220
Logarithmic form, 203
Logarithmic operation, 129
Logic
  easily tested, 288
  incompletely specified, 278
  TTL, 140
Logical convolution, 100
Logical design, 264
Look-up table, 90, 159
LSI, *see* Large-scale integration

M

McCluskey and Quine table method, 264
Magnetic dipole, 255
Majority gates, 270, 272
Majority logic multiplexing, 238
Mask, *see* Template
Matched filtering, 123, 218
Matrix
  Besslich, 279
  binary–Walsh, 266
  C, 206
  factorisation, 58
  Haar, 78
  Hadamard, 24, 28, 30, 166
  permutation, 208
  Rademacher–Walsh, 275
  SAW, 168
  shuffling, 109
  slant, 42, 82

Walsh, 30, 58
  unitary, 119
Mean-square approximation error, 4, 120
Mean-square coefficient, 45
Microcomputer application, 153
Microprocessors, 137
Minimal cover, 278
Minimisation of logic functions, 271
Minimum distance classifier, 222
Minimum distance-to-mean classifier, 226
Minimum integral squared error, 130
Minimum mean-square-error code, 245
Minimum sampling rate, 44
Minterms, 271
  false, 276
  true, 276
Mixed function series, 14, 40
Modulation, 114
  differential pulse-coded, 213
  pulse-coded, 114, 284
Modulo-2 addition, 28, 51, 100, 269
Morse code, 115
MSE, *see* Mean-square approximation error
M-transform, 87
Multi-output system, 287
Multiplexer, 147, 232
  adaptive majority, 240
  block pulse, 233
  digital sequency, 237
  frequency division, 232, 233
  sequency division, 232, 235, 237
  time division, 233
Multiplicative property, 140

N

Natural order, 20, 39
Networks, digital, 268
Non-equivalent symmetry, 272
Non-linear filtering, 129
Non-sinusoidal electromagnetic radiation, 246
Non-sinusoidal waveform, 246
Nyquist interval, 44

O

Odd-harmonic sequency spectrum, 111
Odd-harmonic symmetry, 82

Optimal design, 128
Optimal filter, 218
Ordering, 20, 65, 38
  CAL–SAL, 21, 74
  degree, 38
  dyadic, 20
  Hadamard, *see* Ordering, natural
  Kaczmarz, 19
  natural, 20, 39
  Paley, 20
  Rademacher–Walsh, 22, 265, 268
  rank, 39
  sequency, 15, 19
  Walsh, 19
Orthogonality, 4, 233
Orthogonality error, 140
Orthogonal series, 4
  incomplete, 5
  transformation, 48
Orthonormal, 4
  function series, 4

P

Paley order, 20
Paley-ordered function, 20
Parallel hardware system, 146
Parallel programmable generator, 144
Parity check, 242
Parity digit, 241
Parseval's theorem, 7, 39, 80, 95, 110, 246
Partially symmetric function, 272
Pattern recognition, 202, 220
Pels, *see* Picture elements
Perfect shuffle, 69, 126, 150
Periodogram, 103, 104, 110
  Fourier, 110
  Walsh, 110
Permutation matrix, 208
Personalisation of a PLA, 161
Phase distortion, 233
Phase-invariant transform, 65
Picture elements, 201
Pipeline processors, 152
Pipeline system, 146
PLA, *see* Programmable logic array
Polarity symmetry, 248
  current, 257
Polynomial function, 14
Power-law operation, 129

Power spectra, 70
 frequency-limited, 103
 Haar, 112, 184
 Walsh, 104
Power spectral density, 99
Prime implicant, 277
Prime transform, 162
Processing, 98
Programmable generator, 144
Programmable logic array, 159
Pulse-coded modulation, 114, 213, 284

Q

Quantisation level, 44
Quantisation noise, 154
Quine and McCluskey table method, 264

R

Radar
 carrier-free, 253
 impulse, 247, 251, 253
 into-the-ground impulse, 253
Radar ambiguity function, 252
Rademacher functions, 8, 20, 27, 138, 266
Rademacher–Walsh matrix, 275
Rademacher–Walsh ordering, 22, 265, 268
Rademacher–Walsh series, 8
Rademacher–Walsh spectrum, 267
Rademacher waveform, 8, 257
Radiation
 dipole, 257
 non-sinusoidal, 246
 Walsh, 247
Radiator, 255
 arrays, 258
 travelling wave, 255, 258
Radio sequency filter, 249
Rank order, 39
Read-only structure, 159
Recursive relation, 24
Redundancy, 49
 coding, 241
Reed–Muller alphabet, 244
Reed–Muller code, 244
Reed–Muller expansion, 288
Relative bandwidth, 246

Restoration, image, 214
Reversed butterfly, 63
R transform, 65, 86, 110, 222

S

SAL function, 15, 22, 50
Sample-and-hold amplifier, 122, 149
Sampling interval, 44
Sampling theorem, 44
s.a.1 fault condition, *see* Stuck-at-1 fault condition
s.a.0 fault condition, *see* Stuck-at-0 fault condition
SAW, *see* Surface acoustic wave
Scaler filtering, 120, 123
Semi-infinite interval, 7
Sequency, 15
Sequency division multiplexing, 232
Sequency filtering, 215
Sequency function, 15
Sequency generator, 137
Sequency-ordered fast transform, 59
Sequency-ordered Hadamard matrix, 83
Sequency-ordered transformation, 157
Sequency ordering, 15, 19
Sequency spectrum, 104
Serial programmable generator, 143
Shannon limit, 249
Shift-invariant transformation, 86
 register, 148
 theorem, 18, 55, 99
Shuffle, perfect, 69, 126, 150
Shuffling
 interstage, 148
 matrix, 109
Signal, 3
Signal flow diagram, 61
Signal-to-noise ratio, 126, 217
Sine–cosine functions, 5, 17
Sine–cosine terms, 89
Sine–cosine transform, 89
Sine transform, 206
Single-chip LSI logic, 138
Single-chip Walsh function generator, 158
Sinusoidal carrier, 251
Slant function, 13, 43, 82
Slant–Haar series, 43, 85

Slant matrix, 42, 82
Slant series, 41
Slant transform, 82
Soft errors, 291
Source alphabet, 240
Spectral addition, 275
Spectral analyser, 155
Spectral analysis, 103
Spectral decomposition, 98
Spectral density, 121
Spectral signature, 281
Spectral translation, 270
Spectrum
  odd-harmonic sequency, 111
  power, 70
Spread-spectrum communication, 247
Stacked magnetic dipoles, 256
Statistical approach, 220
Stuck-at-1 fault condition, 286
Stuck-at-0 fault condition, 284, 286
Sub-optimal solution, 48
Sum/differing amplifiers, 149
Sum-of-product function, 159
Superposition, 18
Surface acoustic wave, 167
Surface acoustic wave matrix, 168
Symmetrical function, 87
Symmetrical transform, 50
Symmetry relationship, 15, 55
  odd-harmonic, 82
Synchronisation, 236
Syntactic approach, 220, 226
Synthesis of digital networks, 268

T

Technology-generated system, 45
Television, 210
Template, 215
  matching, 220
Test-pattern generation, 282
Threshold
  adaptive, 127
  criteria, 131
  filtering, 122, 223
  gate, 269
  logic, 269
  logic gates, 265
Time base, 10

Time divisional multiplexing, 232
Time-domain averaging, 126
Time-limited function, 103
Time series, 17
Time shift, 51
  circular, 51, 104
  dyadic, 51, 102
Topographical identification, 223
Totally symmetric function, 271
Transform(s), 145, 267
  algorithms, 48
  CAL–SAL, 74
  chirp-Z, 162
  C matrix, 206
  coding, 210
  compensation, 91
  complex Fourier, 207
  constant-geometry, 50
  conversion, 87
  cosine, 111
  Haar, 75
  Hadamard, 205, 279
  hybrid, 82
  Fourier, 51, 89, 99
    from Walsh, 90
  Karhunen–Loève, 48, 83
  M, 87
  operator, 48
  orthogonal, 48
  phase invariant, 65
  prime, 162
  products, 53
  R, 65, 86, 110, 222
  SAW matrix, 168
  sequency-ordered, 59
  shift-invariant, 86
  sine, 206
  slant, 82
  symmetrical, 50
  two-dimensional, 93, 224
  Walsh, 49
    two-dimensional, 224
Transversal filter, 161
Travelling-wave aerial, 255, 258
True minterm, 276
Truth table, 268
  vertices, 275
TTL logic, 140
Two-dimensional array, 93, 256
Two-dimensional filtering, 127

U

V

W

Two-dimensional Haar function, 217
Two-dimensional transformation, 93
Two-dimensional Walsh function, 202
Two-dimensional Walsh sequency transformation, 224

Unequal error correlation, 245
Unitary transform matrix, 119

Variable word-length coding, 244
Vector filtering, 120
Veitch map, 264
Video sensor, 164
Videotext, 210

Walsh addition theorem, 100
Walsh averaged power spectrum, 110
Walsh carrier, 251
Walsh filtering, 104, 124
Walsh function, 11
Walsh function generator, 138
Walsh function series, 15, 34
   two-dimensional, 204
Walsh–Haar series, 40
Walsh–Hadamard transform, 66
Walsh–Kaczmarz order, 19
Walsh matrix, 30, 58
Walsh matrix order, 20, 39
Walsh order, 19
Walsh periodogram, 110
Walsh power spectrum, 104
Walsh radiation, 247
Walsh sequency order, 15, 19
Walsh series, 36, 49
   derivation, 24
Walsh transform, 49
Walsh transform algorithm, 80
Waveform synthesis, 99, 129
Wiener filtering, 119
Wiener–Khintchine theorem, 104, 108

Z

Zero crossing, 12, 17
Zero-order hold, 45
Zonal bit reversal, 39